358.4483
DOR

FOREST PARK PUBLIC LIBRARY

Air Force One.

32026002895875

W9-AJD-435

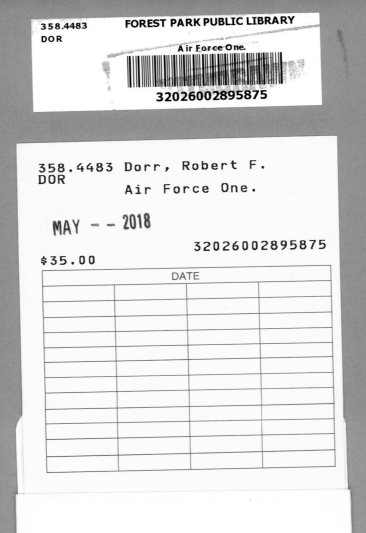

358.4483 Dorr, Robert F.
DOR
         Air Force One.

MAY - - 2018

                              32026002895875
$35.00

DATE			

FOREST PARK PUBLIC LIBRARY

BAKER & TAYLOR

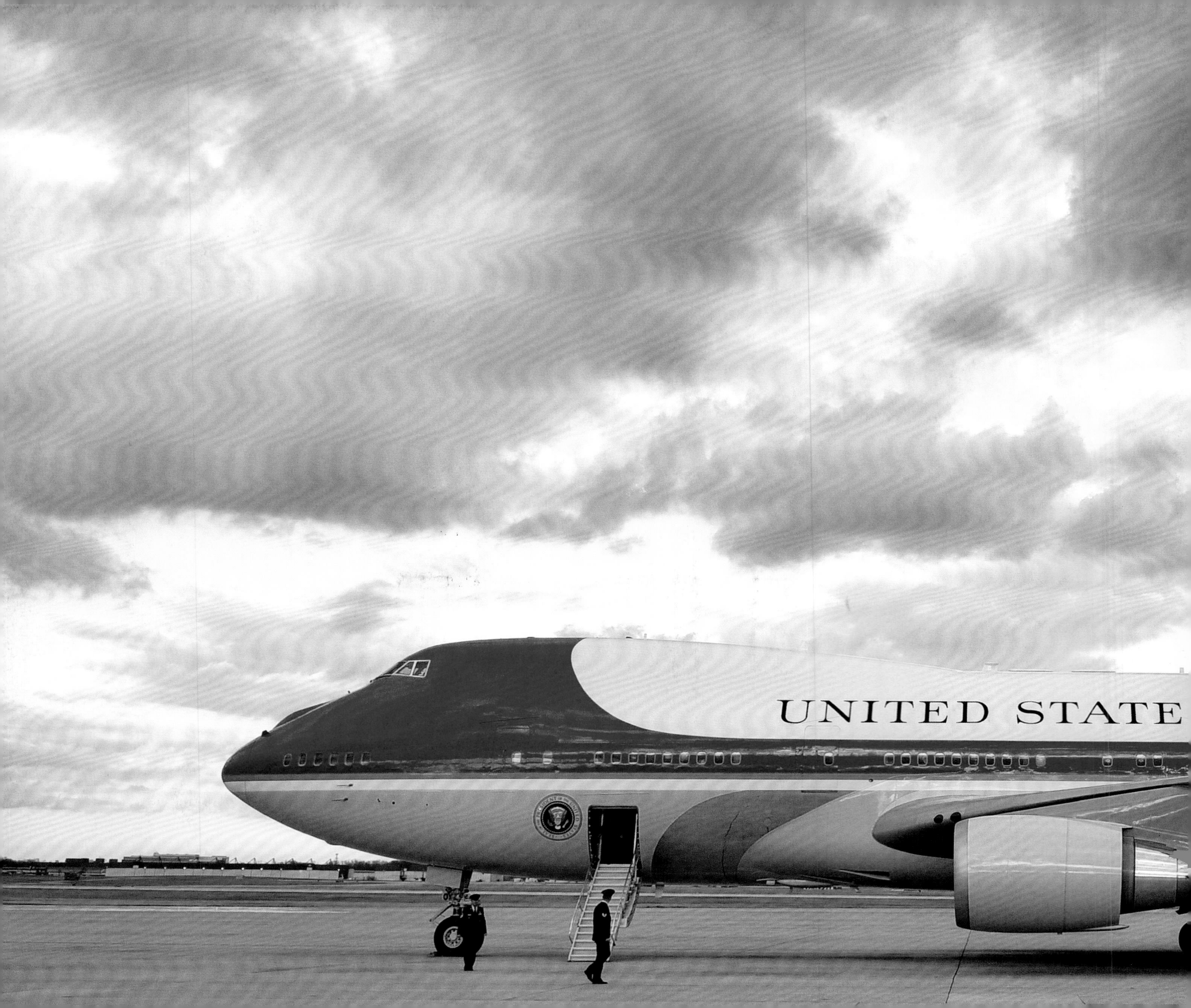
UNITED STATE

AIR FORCE ONE

THE AIRCRAFT OF THE MODERN U.S. PRESIDENCY

ROBERT F. DORR AND NICHOLAS A. VERONICO

motorbooks

FOREST PARK PUBLIC LIBRARY

MAY -- 2018

FOREST PARK, IL

28000

Brimming with creative inspiration, how-to projects, and useful information to enrich your everyday life, Quarto Knows is a favorite destination for those pursuing their interests and passions. Visit our site and dig deeper with our books into your area of interest: Quarto Creates, Quarto Cooks, Quarto Homes, Quarto Lives, Quarto Drives, Quarto Explores, Quarto Gifts, or Quarto Kids.

© 2018 Quarto Publishing Group USA Inc.
Text © 2018 Robert F. Dorr and Nicholas A. Veronico

First published in 2018 by Motorbooks, an imprint of The Quarto Group, 401 Second Avenue North, Suite 310, Minneapolis, MN 55401 USA. Telephone: (612) 344-8100 Fax: (612) 344-8692 www.QuartoKnows.com

All rights reserved. No part of this book may be reproduced in any form without written permission of the copyright owners. All images in this book have been reproduced with the knowledge and prior consent of the artists concerned, and no responsibility is accepted by producer, publisher, or printer for any infringement of copyright or otherwise, arising from the contents of this publication. Every effort has been made to ensure that credits accurately comply with information supplied. We apologize for any inaccuracies that may have occurred and will resolve inaccurate or missing information in a subsequent reprinting of the book.

Motorbooks titles are also available at discount for retail, wholesale, promotional, and bulk purchase. For details, contact the Special Sales Manager by email at specialsales@quarto.com or by mail at The Quarto Group, ttn: Special Sales Manager, 401 Second Avenue North, Suite 310, Minneapolis, MN 55401 USA.

10 9 8 7 6 5 4 3 2 1
ISBN: 978-0-7603-5799-6

Library of Congress Cataloging-in-Publication Data

Names: Dorr, Robert F., author. | Veronico, Nick, 1961- author.
Title: Air Force One : the aircraft of the modern U.S. presidency / by Robert
  F. Dorr and Nicholas A. Veronico.
Description: Minneapolis, Minnesota : Motorbooks, 2018. | Includes
  bibliographical references and index.
Identifiers: LCCN 2017045493 | ISBN 9780760357996 (hardbound)
Subjects: LCSH: Air Force One (Presidential aircraft) |
  Presidents--Protection--United States. |
  Presidents--Transportation--United States. | Boeing 747 (Jet transports)
Classification: LCC TL723 .D67 2018 | DDC 358.4/483--dc23
LC record available at https://lccn.loc.gov/2017045493

Acquiring Editor: Jordan Wiklund
Project Manager: Nyle Vialet
Art Director: James Kegley
Cover Designer: Juicebox Design
Layout: Kim Winscher

Printed in China

MIX
Paper from responsible sources
FSC® C008047

ON THE COVER: (clockwise from upper left): President Obama arrives in Kansas City in 2014. *Mark Reinstein / Contributor / Getty Images*; Vice President Lyndon B. Johnson takes the oath of office aboard Air Force One on November 22nd, 1963, shortly after President John F. Kennedy's assassination as First Lady Jacque Kennedy (right) looks on. *Cecil Stoughton. White House Photographs. John F. Kennedy Presidential Library and Museum, Boston;* Saudi Guards of Honor greet President George W. Bush in Riyadh outside Air Force One on May 16, 2008. *MANDEL NGAN / AFP / Getty Images*; Executive Orders await President Obama's signature aboard Air Force One. *JEWEL SAMAD / AFP / Getty Images*; VC-118A (serial number 53-3240) is today preserved at the Pima Air and Space Museum in Tucson, Arizona. *Rick Turner*

ON THE BACK COVER: This unique photo depicts both VC-25A Air Force Ones, displayed for the public at the Joint Base Andrews open house in 2005. Both aircraft taxied out and took off, much to the crowd's delight. *Ken Kula*

ON THE TITLE PAGE: Air Force One is seen on the tarmac of Andrews Air Force Base in Maryland on March 19, 2017. *MANDEL NGAN/AFP/Getty Images*

ON THE ENDPAPERS: VC-137B (serial number 58-6971) is shown approaching Andrews Air Force Base. *Jim Hawkins collection*

Contents

Dedication and Acknowledgments

Air Force One is the story of a plane that belongs to a nation and a history of presidential travel. Any errors in this book are the fault of the authors. Many individuals helped to make this book possible.

A gracious former first lady, Nancy Reagan, gave this author a few minutes to share memories of Air Force One.

I have a special gratitude to former Air Force Chief of Staff General Ronald Fogleman, former Secretary of the Air Force F. Whitten Peters, and former Secretary of the Air Force Dr. James Roche for their personal encouragement of my writing. I owe a similar debt to two successive commanders of the 89th Airlift Wing: Major General James Hawkins and Colonel Glenn Spears.

I also want to thank the following members of the 89th Airlift Wing: Steve Anderson, John Atkins, Richard Balfour, Keith Blades, Dana Carroll, Bruce Christensen, Susan Coke, Rick Corral, Gene Dickson, Howie Franklin, Dennis Fritz, Jeff Gay, Vaughn Gonzales, John Haigh, Jennifer Hider, David Huxsoll, Silkya J. Irizarry, Bobby Jones, Nathan Jones, Susan Koch, Valerie Martindale, Carolyn McPartlin, Kevin McQuay, Kris Meyle, Allison Miller, Doug Normour, Heidi Ostreich, Richard Parker, Susan Richardson, Bob Ronck, Dave Rossner, Bel Serocki, Robert Shaffer, Dave Sims, Donnell Smith, Pamela Varon, Nancy Vetere, Armando Visitacion, Nicky Williams, Coennie Woods, Ronald Zaremba, and Michael Zepf.

In addition, I would like to thank the following people for their help: Kathy Bienfang, Colin Clark, Bill Crimmins, John W. Darr, Greg L. Davis, Clint Downing, Jerry Geer, Jim Goodall, John Gourley, Sunil Gupta, Joseph G. Handelman, Alex Hrapunov, Dennis R. Jenkins, Tom Kaminski, Craig Kaston, Jim Kippen, Patrick Martin, David W. Menard, Robert C. Mikesh, David Ostrowski, Jeffrey Rhodes, Barry Roop, Charles Taylor, and Jack Valenti.

Special thanks go to Robert C. Mikesh, the leading American authority on presidential aircraft, for permission to adapt material from his article in the Summer 1963 issue of the *American Aviation Historical Society* journal for use in Chapter 2 of this book. Another special nod, too, to Kirsten Tedesco, Stephanie Mitchell, and James Stemm at the Pima Air and Space Museum in Tucson, Arizona. Some of the material on presidential helicopter travel comes from the government publication *Naval Aviation News*.

Air Force One is dedicated to Marc Reid, who is finally able to straighten up and fly right.

Robert F. Dorr
Oakton, Virginia

For the updated edition, special thanks to Ian Abbott, Mark Aldrich, Manny Antimisiaris, Stewart Bailey, Brian Baum, Andrew Boehly at the Pima Air and Space Museum, Roger Cain, Michael Carter, Ron Collins, Janice Davis at the Harry S. Truman Library, Yvonne Densem, Jim Dunn, Chris Farinha, Hank Frakes, Alan Griffith, Joe Handelman, Mike Henniger, Dennis Jenkins, Tom Kaminski, Joe Kates, Ken Kula, William T. Larkins, Harry Lloyd, Mark Nankivil, Robert Nishimura, Jon Proctor, Frank Prokup, Joe Pruzzo at the Castle Air Museum, Ken Rice, Tony Rocha, Ron Strong, Spike Tellier, Andrea Tibbotts, Rick Turner, Betty Veronico, Tony Veronico, and Jordan Wiklund.

Nicholas A. Veronico
San Carlos, California

The Flying White House

At Joint Base Andrews in Maryland, outside the nation's capital, the president of the United States is about to take a trip. The sky giant that will carry the president is a marvel of engineering. Just now, the plane is indoors—the president will climb aboard while it is still inside its hangar—but to those who can see it, the aircraft catches the eye and holds the gaze.

Many Americans would recognize it as a Boeing 747, the jumbo jet that opened up air travel to the everyday citizen. This one is a military version, known in jargon as a VC-25. But there is nothing ordinary about this particular aircraft and nothing obscure about the name by which most of the world knows it. To the press, public, moviegoers, and air enthusiasts—even to the people who work on it—it is called Air Force One.

FLYING WHITE HOUSE

Air Force One glistens, its blue-and-white exterior polished to a bright shine. On the inside, its engines, avionics, and instruments have been tweaked to perfection. There is probably not a 747 in the world that looks so pristine or is in such flawless running order. Air Force One is maintained by the best men and women the nation can recruit for the job. As one of them says in a quick aside, good naturedly and with considerable pride, "You do not make a mistake when working on Air Force One."

The president of the United States is the most important customer of the air force's 89th Airlift Wing, the resident military unit at Andrews. The wing provides transport to members of Congress, the Cabinet, and sometimes kings and kingmakers, but the president receives very special treatment. He, and at some point in the future, she, needs to be able to communicate with anyone in the world at any time. The president needs to be able to travel to any location on short notice with the ability to change plans with no notice. When he heads off to some distant capital to confer with an international leader, the president needs to arrive safe, comfortable, refreshed, and ready.

PRESIDENTIAL ARRIVAL

President Barack Obama arrives at Joint Base Andrews, Maryland, on March 29, 2016, by motorcade. He departed aboard Air Force One, which is an 89th Airlift Wing VC-25A aircraft modified for presidential support. The Presidential Airlift Group operates two VC-25s and supports the president with a full complement of selectively assigned pilots, navigators, flight engineers, communications system operators, flight attendants, and maintainers. The VC-25A is specially equipped to meet the president's needs with accommodations that include an executive suite with a stateroom, office, conference room, and dining room. *US Air Force photo by Senior Master Sgt. Kevin Wallace*

So what is it like when the president arrives at Andrews to climb aboard his flying magic carpet? It is a sight to behold, and not solely because of the magnificence of the chief executive's four-engine, 833,000-pound Boeing aircraft.

The chosen few among the press who cover the president are asked to arrive at Andrews an hour before his departure. En route to the fenced-in outdoor area set aside for the press, everyone stops to be searched. Identification is double checked. A Secret Service agent inspects each individual rigorously using a handheld metal detector. Photographers must lay their cameras on the ground and have them sniffed by a Secret Service sentry dog. Meanwhile, the crew of Air Force One readies the giant jet inside the presidential hangar—it is almost time to fly.

While these preparations go on, the Secret Service may order a "quiet period" (if the president is planning to speak before departure) or a "ramp freeze" (halting all aircraft movement on the west side of the Andrews's runway 01/19) but is more likely to order both. Quiet means quiet. No other aircraft can taxi on the west side or make an afterburner takeoff during a "quiet period."

Before the president arrives or Air Force One taxis out from its hangar, a vehicle from the 89th Wing's base

After traveling across the District of Columbia by helicopter, President Barack Obama steps off Marine One and is greeted by two marines and Col. J. C. Millard, 89th Airlift Wing commander, at Joint Base Andrews on July 15, 2015. Obama departed Andrews aboard an 89th Airlift Wing C-32A, call sign Air Force One.
US Air Force photos by Senior Master Sgt. Kevin Wallace

operations center inspects the entire flight line west of the runway for foreign object debris (FOD), fallen objects of any size that might be ingested by the 11-foot-wide intakes of Air Force One's four General Electric F103-GE-180 turbofan engines. The flight line is routinely policed for FOD, but an extra inspection precedes every departure by the chief executive.

If the president is coming from the White House by motorcade, air force personnel will first observe his vehicles at the Andrews main gate just off the Washington Beltway, alias Interstate 495. An advance party will have reached the gate first, and the president will be whisked through. The presidential motorcade then proceeds east on Westover Street, then south on Arnold toward the Air Force One hangar, never slowing, never pausing.

Almost always, there at least two identical limousines in the motorcade, one of them a decoy. In addition, there are at least two gloss black Chevrolet Suburban "war wagons" carrying Secret Service agents and their weaponry. While the presidential limousine goes directly to the VIP area (if a speech is planned) or to the hangar (if not), two sharpshooters from the Executive Protection Agency (the uniformed arm of the Secret Service) are at the ready on the ledge outside the Andrews control tower—about 1,000 feet from the hangar—with 7mm Magnum sniper rifles with scopes inside aluminum gun cases. Additional Secret Service officers occupy other key positions surrounding the area of presidential activity.

In approximately 90 percent of presidential departures, the chief executive does not arrive by motorcade. The first sign of his appearance is the throb of the presidential helicopter, Marine One, a Sikorsky VH-3H or VH-60N. Either way, the destination is the hangar where Air Force One awaits.

Inside the hangar (tall enough to accommodate the looming 63-foot 5-inch tail of a 747) two VC-25As are awaiting their famous passenger, one of the aircraft serving as a backup. There is always a backup. If one of the 747s is temporarily out of action for any reason, including periodic depot maintenance, some other transport from the 89th Airlift Wing serves as the backup instead. During routine operations in the United States, the 89th usually does not provide a decoy aircraft. Overseas, however, it may. Up to the actual time of boarding, only a handful of people know which aircraft the president will board.

PRESIDENTIAL PLANE

The VC-25A, as the air force calls its Boeing 747-200 (Air Force One, to most of us), is a military, VIP version of the plane that changed the world. According to a survey of aviation analysts at the Los Angeles–based Ralston Institute, as recently as 1970 only 25 percent of Americans had ever been inside an aircraft. Most Americans traveled by car, bus, or train. A seat on an airliner was the special province of the privileged few. At that time, most airline passengers were business people or well off. The Boeing 747 changed all that. Introduced to commercial airline service in 1970, the 747 doubled passenger capacity overnight. The wide-body, or jumbo, era had arrived. Within a few years, passenger loads on the nation's interstate buses began to decline. Those on the airlines began to skyrocket. It was size that mattered, not speed. The idea of an American supersonic transport did not last beyond 1970, but the concept of a huge aircraft, able to carry an unprecedented number of people, remains alive today.

The version of the 747 with the highest-density seating on today's airline routes, the 747-400, carries up to 624 people. The 747 assigned to carry the president of the United States exists for only one passenger, but it often carries as few as two dozen and rarely more than sixty people.

Although airliners with two engines and a two-person flight crew routinely circle the globe today, Air Force One was ordered at a time when the Pentagon's air staff insisted that any plane carrying the president must have four engines and a four-person flight crew (pilot, copilot, navigator, and flight engineer). So Air Force One pilots have plenty of help, but their job is crucial, and the system allows them no room for mistakes.

PREFLIGHT BRIEFING

Long before the president heads for Andrews to begin his trip, the aircraft commander (that is, the pilot in command) gathers the entire flight crew, including communications operators and flight attendants, around a conference table in the Presidential Airlift Group (formerly the Presidential Pilot's Office). Using charts and maps, the aircraft commander informs everyone where they're going and how they'll get there. Then the aircraft commander turns the floor over to an intelligence officer who briefs the crew on their destination. It may be Chicago, and this portion of the briefing may consist of nothing more than a description of a hotel. Or the destination may be Vilnius, Lithuania, and the intelligence briefing will include the prospects of being robbed by pickpocket or attacked by terrorists.

The intelligence officer who, as of this writing, provides the intelligence briefing to the crew probably needs to know about more disparate subjects than anyone else in the air force. The intelligence officer's knowledge ranges from the performance of Russian-built MiG fighters to the personality of the Lithuanian prime minister. Unlike the rest of the Air Force One crew, this person has access to "all source intelligence," meaning that he is cleared for the information that results from communication intelligence and satellite reconnaissance. Contrary to myth, Air Force One is almost never escorted by friendly fighters, but the intelligence officer must tell the crew whether anybody else's fighters will be in the area.

The intelligence briefing will range from a few minutes to an hour, depending on the destination, the level of threat (if any) confronting Air Force One, and the amount of intelligence available on the mission. It should be emphasized that the president's security, and the intelligence related to it, remains the job of the Secret Service and is handled separately from the preparations by the Air Force One crew for its mission.

After the intelligence brief, the aircraft commander presents the flight plan. At this juncture, the airborne communications systems operator (ACSO) reviews the communication needs for the flight. It is impossible to

President Donald Trump and his wife Melania Trump board one of two VC-25As that serve as Air Force One. The 89th Airlift Wing provides both the VC-25A and C-32 (a military version of the Boeing 757) for use by the president and other high-ranking government officials. *Photo by Alvaro Padilla Bengoa/Anadolu Agency/Getty Images*

exaggerate the importance of the ACSO—still called a radio operator in everyday conversation—to the mission of the 747.

Air Force One has crew positions for three ACSOs and almost always carries a fourth in reserve. The communications center is located in the stretched upper deck of the aircraft, used on the commercial version to carry passengers. Today aboard Air Force One, the ACSO will operate radios, but he must also enable the president and other passengers to use telephones, email, and the entertainment system. After briefing with the crew, the ACSO will also brief with the Secret Service. While Air Force One is in flight, the ACSO will be the eyes and ears of the super-secret White House Communications Agency, which keeps the president in touch and ready to react to any crisis.

Like the other crewmembers, Air Force One's flight attendants have also been preparing. Flight attendants begin planning forty-eight hours prior to a presidential mission. From one end of the resplendent blue-and-white VC-25A to the other, the flight attendants are responsible for the cleanliness and orderly functioning of 4,000 square feet of cabin space. While the flight is being planned, they stock groceries (purchased anonymously at a constantly shifting roster of retail outlets), spruce up furnishings and furniture, and even make the president's bed. There are two presidential bunks, and one is always kept ready for the chief executive, should he choose to use it. As one flight attendant told the author, "Whether it's a day trip or a summit trip, everything must work right."

Air Force One taxis as it prepares to fly to Selma, Alabama, from Joint Base Andrews, Maryland, on March 7, 2015. President Barack Obama, First Lady Michelle Obama, and their teenage daughters, Malia and Sasha, traveled south to mark the fiftieth anniversary of Bloody Sunday in Selma, where state troopers violently attacked a peaceful civil rights march on the Edmund Pettus Bridge on March 7, 1965. *US Air Force photo by Senior Master Sgt. Kevin Wallace*

DEPARTURE

The pilots, flight crew, and the remainder of Air Force One's crew are inside the airplane, buttoning up and preparing to go long before the chief executive's motorcade arrives. The last step consists of the aircraft commander and copilot running through the before-start checklist, with one uttering the "challenge" (in the tone of a question) and the other giving the "response":

Q: Departure briefing?
A: Complete.
Q: FMS [flight management systems], radios?
A: Programmed, set, verified.
Q: IRSs [inertial reference systems]?
A: Nav [navigation] aligned.
Q: Hydraulic demand pumps?
A: Number 1 auto, number 4 aux [auxiliary].
Q: Oil quantity?
A: Normal.

The pilots inform the crew and those monitoring their preparations from outside the aircraft that they are ready to take off. They obtain clearance to taxi. The pilots receive a "block time" for departure. They will attempt to meet that deadline within ten seconds. Usually, everything goes like clockwork.

But first, they wait.

The last person aboard before the president is a military officer (an air force, army, or marine colonel, or a coast guard or navy captain) carrying the "football," the package containing the codes that enable the president to launch a nuclear strike.

VC-25A (serial number 82-9000), serving as Air Force One, departs from snow-covered Joint Base Andrews, Maryland, with President Obama on board. The VC-25A is similar to the Boeing 747-200 in that it is powered by four General Electric CF6-80C2B1 turbofan engines, rated at 56,700 pounds each, that propel the jumbo jet through the sky at a maximum speed of Mach 0.84, with a range of 6,735 miles. The VC-25A's maximum gross weight is 833,000 pounds. The aircraft is staffed by a crew of 26 and can carry 102 passengers. *US Air Force photos by Senior Master Sgt. Kevin Wallace*

continued on page 18

ABOVE: Air Force One VC-25A (serial number 82-8000) touches down at Long Beach Airport, California, approximately 30 miles south of downtown Los Angeles, on March 18, 2009. President Obama then traveled by motorcade, rather than helicopter, to the Orange County fairgrounds in Costa Mesa, where he held a town hall meeting during a two-day visit to the area. *Michael Carter/Aero Pacific Images*

LEFT: A US Marine stands at ease next to a VH-60N White Hawk executive transport helicopter as Air Force One lands at Buckley Air Force Base, Colorado, on April 24, 2012. President Obama visited the University of Colorado Boulder to talk to students about a looming spike in student loan interest rates. The president flew from Buckley to Boulder in a VH-60N rather than make the 40-mile drive by motorcade. *Technical Sgt. Wolfram M. Stumpf*

continued from page 15

The job is rotated among officers of the five military service branches. The protocol that places the "football" so close to the president and gives the device its name has solidified into ritual. By law, only the president or the secretary of defense can authorize the release of nuclear weapons. The device is simply a large briefcase, but parlance requires that it be called the "football" and nothing else.

If the president's party arrives at Air Force One on time, he and his fellow travelers come aboard exactly as the pilots complete their before-flight checklist. Air Force One's tires will begin to turn, almost always, exactly at "block time." Otherwise, the pilots and crew will wait on their passengers. It is the opposite of what happens on commercial airlines. Because of his importance, and also for security reasons, the boarding of the president is the last thing that happens before Air Force One begins to roll.

Air Force One taxies much faster than airliners usually do, its crew secure in the knowledge that no aircraft or vehicle will be in their way. The pilot in command obtains takeoff clearance while rolling, and the two pilots and the flight engineer complete the before-takeoff checklist while in motion. Most of the time, the VC-25A rounds the turn onto the active runway and begins its takeoff roll without hesitation. With the enormous power of its four engines unleashed—the equivalent of two railway locomotives, or enough thrust to propel the VC-25A straight up—the mighty aircraft lifts into the sky and begins its flight, the airways ahead being cleared as it climbs.

Before the president's aircraft departs, a fleet of other aircraft is well on the way, or already at the destination, with necessary travel items. A presidential limousine often goes with the president, carried by an air force cargo plane.

President George W. Bush and First Lady Laura Bush wave from the steps of Air Force One while watched by the president's military aide who holds the briefcase called 'the football,' which contains the nuclear codes to launch missiles. *Photo by David Hume Kennerly/Getty Images*

BOTTOM LEFT: Staff Sgt. Matthew Smith, a 447th Expeditionary Security Forces Squadron (ESFS) member, guards Air Force One on April 7, 2009. Members of the 447th ESFS protected Air Force One while President Barack Obama spoke to a crowd of nearly 1,500 service members, government contractors, and civilians at Al Faw Palace, Camp Victory, Iraq, during an unannounced visit to the war-torn nation. *Staff Sgt. Amanda Currier*

BOTTOM RIGHT: Staff Sgt. Donald Maddox, a 447th Expeditionary Logistics Readiness Squadron (ELRS) fuel distribution technician, refuels Air Force One on April 7, 2009, at Camp Victory, Iraq, while Master Sgt. Joel Sutton, 447th ELRS fuels manager, and an Air Force One crew chief stand ready to assist. *Staff Sgt. Amanda Currier*

A duplicate limousine and Secret Service vehicles are also carried to the destination by air force cargo planes. Everything is in place when the chief executive arrives.

ON ARRIVAL

Speed—not haste, but measured, prudent, and deliberate speed—is a factor in the operation of Air Force One at all times. A high-speed taxi is often vital to the plan for the day, not only on departure but also on arrival as well. When President Ronald Reagan lay on the operating table after being shot by a would-be assassin on March 30, 1981, Air Force Two happened to be bringing Vice President George Bush home from a trip. No one knew whether there was a further threat to US leaders or whether, at any instant, Reagan might die, making Bush president. Bush's plane taxied at a rapid pace into the hangar, and the vice president did not emerge from the aircraft until the hangar had been shut down and the aircraft surrounded by security.

Even a landing in normal times triggers a flurry of activity. The military officer with the football hits the ground first and hands off his precious package to his successor. The president prepares to face his public, inevitably before a microphone. When he steps off the aircraft, a general or colonel greets him at the base of the airstairs. If the arrival site is Andrews, the greeter is usually the 89th Wing commander.

PREVIOUS PAGES: A convoy gathers around Air Force One after it landed on July 13, 2017 at Paris' Orly airport with US President Donald Trump and First lady Melania onboard.

TOP: A Marine Helicopter Squadron One (HMX-1) VH-60 departs Marine Corps Air Station (MCAS) Iwakuni, Japan, on May 27, 2016, after transporting President Obama. The chief executive had visited MCAS Iwakuni and spoken with service members and their families after the Ise-Shima Summit. *US Marine Corps photo by Cpl. Nathan Wicks*

BOTTOM: President Barack Obama salutes two US Marines as he boards Marine One at Buckley Air Force Base, Colorado, on April 24, 2012, preparing to return to Air Force One for the next leg of his trip. *Technical Sgt. Wolfram M. Stumpf*

OPPOSITE: A pair of HMX-1 Sikorsky VH-60N White Hawk executive transport helicopters, one serving as Marine One and the second as a decoy Marine One, departs Buckley Air Force Base, Colorado, on April 24, 2012. In addition to Marine One, there can be as many as three additional helicopters serving as decoys when the president is being transported. *Technical Sgt. Wolfram M. Stumpf*

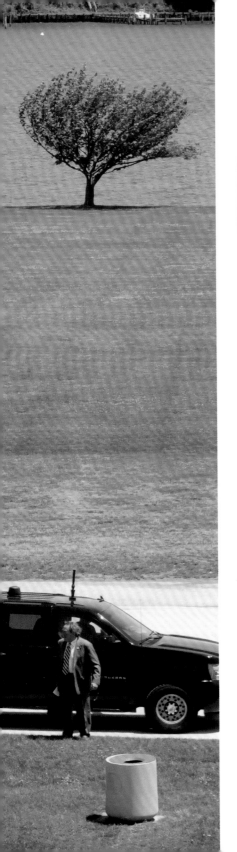

ABOVE: President Donald Trump salutes US Marines from the presidential guard detail during his first overseas speech to troops at Naval Air Station Sigonella, Italy, on May 27, 2017. Trump traveled to Sicily to attend the G7 Summit and meet with other world leaders, including Italian Prime Minister Paolo Gentiloni. Naval Air Station Sigonella is strategically located near Catania at the eastern end of Sicily and supports an array of US Navy and Marine Corps operations in the region. *US Marine Corps photo by Sgt. Samuel Guerra*

LEFT: When traveling by Marine One helicopter, the president and vice president are met at their destination by Secret Service vehicles that will form the motorcade. Here, Vice President Joe Biden arrives at Annapolis, Maryland, to give the graduation speech at the US Naval Academy. *J. G. Handelman*

CHAPTER 2 TWO

Propellers for the President

Theodore Roosevelt may have been the most adventurous American president. The man who led his Rough Riders in the courageous charge at Cuba's San Juan Heights during the Spanish-American War continued to show his audacious nature by seeking out friendship with Wilbur and Orville Wright. He encouraged the pioneering brothers by proclaiming that the airplane had "a great future."

While participating in the Missouri State Republican campaign in Saint Louis, Missouri, on October 11, 1910, Roosevelt was invited to fly in a Wright Type B biplane with pilot Arch Hoxsey. As the twenty-sixth president from 1901 to 1909, Roosevelt was not in office when he made the flight, but by sitting in a shuddering biplane made of wood and fabric, he defied the accepted wisdom of what was safe for a president and what was not. Critics said that Roosevelt was too adventurous and that he was pushing things too far.

Surviving motion picture footage exists of Roosevelt arriving at the Kinloch aviation field (accompanied by Herbert S. Hadley, governor of Missouri), entering the passenger seat, flying, and descending to join his waiting party. Roosevelt later commented to a *New York Times* reporter, "You know I didn't intend to do it, but when I saw the thing there, I could not resist it." No man was a better candidate to become the first president to fly.

Roosevelt's successors as chief executive—William Howard Taft (1909–1913), Woodrow Wilson (1913–1921), Warren Harding (1921–1923), Calvin Coolidge (1923–1929), and Herbert Hoover (1929–1933)—walked, rode in automobiles, or took the train while in office. There is no record, in fact, that any of these men flew in an aircraft while they served as president of the United States, although Hoover certainly did so after leaving office.

OPPOSITE: Pan Am's Boeing 314 *Dixie Clipper* was the first aircraft to transport a president while in office, flying Franklin D. Roosevelt from the United States to Africa on the over-water portion of the trip to Casablanca, French Morocco. Here, Roosevelt, British Prime Minister Winston Churchill, and French Generals Charles de Gaulle and Henri Giraud charted the course of the war against Nazi Germany with the demand for an unconditional surrender. *Ed Davies collection*

Before he became the thirty-second president (1933–1945), Franklin D. Roosevelt had several occasions to fly. The most famous occurred in July 1932 when, already chosen as his party's nominee for the presidency, FDR flew from Albany, New York, to Chicago, Illinois, to attend the Democratic convention in an American Airways, Inc., Ford Tri-Motor registered as NC415H. With its mostly metal structure and corrugated fuselage skin, the ungainly Tri-Motor was known as the "Tin Lizzie," but the name was given affectionately. It was a reliable flying machine and performed its duty for FDR without a hitch.

FDR eventually became the first president to fly while in office; the trip occurred at the height of World War II. Many in the press and public still regarded aviation as dangerous, but there was no rule, written or unwritten, about the chief executive flying in an aircraft. The world was on the brink of change.

THREE PLANES FOR FDR

There is evidence that a navy Douglas RD-2 Dolphin was assigned to presidential support duties in 1936 during FDR's time in office. The nation's admirals might have seen that presidential air travel was inevitable and may have had in

Having been selected as his party's nominee for president, Franklin D. Roosevelt made a flight in an American Airlines Ford Tri-Motor similar to the one shown here. The interior was cramped with seating for eight to ten guests, depending upon interior configuration, and a flight crew of two. The Tri-Motor was not fast, having a top speed of 199 miles per hour, but the aircraft was reliable. *William T. Larkins collection*

The US Navy used a Douglas RD-2 Dolphin, identical to the US Coast Guard RD-2 Adhara shown here, to deliver mail to President Roosevelt while he was on the presidential yacht *Potomac* in 1936. Should Roosevelt have needed to be transported from the yacht back to Washington, DC, the navy's RD-2 was on standby. *William T. Larkins collection*

mind launching an early bid to land the job of carrying the chief executive.

No prettier than the Ford Tri-Motor but as practical as they came, the Dolphin was a popular high-wing, twin-engine, monoplane amphibian powered by two 350-horsepower Wright R-975-3 radial engines and capable of a maximum speed of 140 miles per hour. It was a popular aircraft in the 1930s with the army, navy, and coast guard.

RD-2 Dolphin Bureau of Aeronautics serial number (Buno.) 9347 was the subject of a letter from the commander of Naval Air Station Anacostia in Washington, DC, to the chief of the navy's Bureau of Aeronautics, reading in part, "Although not exclusively used for the President and the Sec[retary] of the Navy, this plane was set aside to carry the mail to the President when he was embarked in the Potomac [River, aboard the presidential yacht] and it was used by the Assistant Sec[retary] of the Navy, the late Colonel H. L. Roosevelt, for official transportation. It is very necessary to have a plane of this type on this station all times for transportation of mail and guests of the President."

Navy records are ambiguous as to whether the service ever intended the Dolphin specifically to carry the chief executive. Undoubtedly, the navy must have hoped it would receive the honor of transporting the president. If the Dolphin was intended in part for that purpose, it became the first aircraft ever to draw such an assignment—but the president never flew in it.

Additionally, a Curtiss YC-30 Condor twin-engine biplane stationed at Bolling Field in Washington, DC, in the late 1930s was believed by many to be earmarked as a presidential aircraft. No record confirming this has ever been found, and Roosevelt did not fly in the Condor either.

Lastly, the *Guess Where II*—the name a play on Guess Where To?—a transportation version of the consolidated B-24D Liberator four-engine heavy bomber known as a C-87A, was assigned for presidential travel for a period of more than one year. During this time, it completed several extensive trips. One of these was an around-the-world journey begun on July 25, 1943, carrying five senators on what would be known today as a fact-finding tour. *Guess Where II* also carried First Lady Eleanor Roosevelt but never her husband. Army Air Forces flyers were severely disappointed that their plans to transport the boss in the C-87A never materialized.

FIRST PRESIDENTIAL FLIGHT

Despite the objections of many in the Secret Service and in the executive mansion, Roosevelt mustered his great sense of historical drama and used an aircraft to fly from Miami, Florida, to Casablanca, Morocco, in January 1943 to attend a meeting with Prime Minister Winston Churchill and other Allied leaders to plot strategy for the war. Although plans were forming for a military squadron to transport the president by air, Roosevelt found himself journeying aboard a civilian plane. The Secret Service had ruled out travel by sea, leery of the threat from German submarines, but the protectors of the chief executive were not happy about Roosevelt flying.

The attitude of at least one observer was summed up by columnist Christopher Wayne of the *Washington, D.C., Evening Star*, who wrote in a January 1943 editorial piece that "aviation is demanding and unforgiving. It is still not clear that the obvious advantages of air travel carry greater weight than the peril incurred by risking the life of a president on a journey through the sky."

Flying Boat Trip

Roosevelt took off from Dinner Key seaplane base in Miami on January 11, 1943, as a passenger in a Boeing 314 Clipper. The aircraft wore the civil registry number NC18605 and was named the *Dixie Clipper*. The plane was also assigned the US Navy Buno. 48226 and constructor's number 1992.

Roosevelt traveled on the presidential railcar, the *Ferdinand Magellan*, from Washington, DC, to Miami, where he boarded the 314 Clipper, a long-range, four-engine flying boat used for transatlantic flights before World War II. For the president's comfort, a double bed was installed. The huge flying boat was one of several operated for the navy by Pan American Airways (known as PAA—the nickname Pan Am still lay in the future). Lt. Howard M. Cone of the US Navy Reserve was the pilot during this first presidential flight.

At the time, it was not only controversial for a president to fly, but it was also controversial for a president to leave American soil. None had previously done so during wartime, and no president since Abraham Lincoln had ever traveled to a war zone.

Roosevelt's flying trip was enough to test anyone's stamina. With eight other passengers accompanying the president, pilot Cone embarked on a journey consisting of three segments, beginning with a 1,633-mile flight to Port of Spain, Trinidad. The next day, the silver, triple-tailed Clipper continued on a 1,227-mile leg for Belém, Brazil, covering the distance in about eight hours. The following day, the Dixie Clipper, again piloted by Cone, flew the longest leg of this expedition, covering the 2,500 miles to Bathurst, British Gambia, in West Africa (known today Banjul, Gambia).

Presidential Plane Number 2

On the heels of Clipper pilot Cone, the second pilot to provide transportation to Roosevelt was Otis F. Bryan, a major in the army reserve and a vice president of Trans World Airlines (TWA). Bryan was waiting in Bathurst with a Douglas C-54 Skymaster, which was operated by TWA under a wartime contract to the government.

As a C-54 passenger, Roosevelt flew 1,500 miles to Casablanca, Morocco, for one of several crucial conferences with Allied leaders. He arrived on January 14, 1943. After spending a fortnight in North Africa, Roosevelt returned to Washington by retracing his path aboard the same two aircraft.

Continued on page 34

President Roosevelt celebrated his birthday on January 30, 1943, during the transatlantic crossing home from the Casablanca Conference on board the *Dixie Clipper*. Joining the president, from left to right, were Fleet Adm. William D. Leahy (chief of staff to the commander in chief), Roosevelt, Harry L. Hopkins (secretary of commerce during the New Deal and subsequently Roosevelt's chief diplomatic advisor during World War II), and the plane's pilot, Lt. Howard M. Cone (US Naval Reserve). *Courtesy National Museum of the US Air Force*

PRESIDENTIAL C-87A LIBERATOR
GUESS WHERE II

The distinction of being the first aircraft customized for presidential travel belongs to the military transport version of the Consolidated B-24D Liberator four-engine heavy bomber. On June 2, 1943, this C-87A Liberator Express (41-24159) was accepted by the 503rd Army Air Base Unit at Washington National Airport. The first presidential pilot, Maj. (later Col.) Henry T. "Hank" Myers, dubbed the unit the "Brass Hat Squadron" and named the aircraft *Guess Where II*. An important consideration in the choice of aircraft was a long range to avoid the security problems that might arise with frequent stopovers. The C-87A offered a rather extraordinary range of 2,900 miles.

According to retired Maj. Robert C. Mikesh, the leading authority on presidential aircraft, Myers intended the name of the C-87A to be a word play on the question "Guess where to?" There was never an aircraft named *Guess*

Where I. It is a myth that the name came from Roosevelt relating to his famous quote about the April 1942 bombing raid on Japan by Lt. Col. James Doolittle's B-25 Mitchell raiders. The B-25s had taken off from an aircraft carrier in the western Pacific, but when Roosevelt was asked their point of departure, he first challenged a reporter to guess where, then replied with alacrity, "Shangri-la." The latter was the fictitious redoubt in James Hilton's novel *Lost Horizon*; it was the president's way of saying that the reporters could keep right on guessing.

Wearing standard army olive-drab paint with a light gray underside, *Guess Where II* boasted four compartments similar to the sleeper compartments of railroad trains of the era. It could accommodate nine VIP passengers with overnight sleeping arrangements or twenty passengers by day. The C-87A also had two lavatories and a full galley.

The army acquired 280 C-87 and 5 C-87A models for transport duty, all built by Consolidated in Fort Worth, Texas. All were powered by four 1,200-horsepower Pratt & Whitney R-1830-43 Twin Wasp eighteen-cylinder turbocharged radial piston engines. All had a cruising speed of 200 miles per hour and could probably work their way up to 290 miles per hour for brief periods under the right conditions. Air Transport Command pilot Ernest K. Gann was one of many who was not enthusiastic about these transports made from bombers. "The assembly of parts known collectively as a C-87 will never replace the airplane," Gann opined. "They were an evil, bastard contraption, nothing like the relatively efficient B-24 except in appearance . . . the C-87s could not carry enough ice to chill a highball."

Apparently, *Guess Where II* was ready by the time Roosevelt made his second overseas trip in November 1943 to meet British Prime Minister Winston Churchill and the Soviet Union's Premier Josef Stalin at Tehran and Cairo. But Myers, the

Consolidated C-87A Liberator (serial number 41-24159) wore the name *Gulliver II* early in its career with the Air Transport Command. The aircraft's interior was converted to a sleeper configuration for President Franklin D. Roosevelt, but he never used the Liberator. Renamed *Guess Where II*, First Lady Eleanor Roosevelt made extensive use of the transport, visiting troops and making a Latin and South American goodwill tour. *US Army Air Forces via Alan Griffith*

C-87A, and the presidential crew were not tapped for the job. Reportedly, the Secret Service had a view of the C-87A not dissimilar to Gann's. "A firetrap," one critic is supposed to have argued. As recounted in the accompanying text, Roosevelt flew a different aircraft on the journey.

The Secret Service and the army never permitted *Guess Where II* to perform the presidential mission for which it had been built. Nevertheless, serving with the Brass Hats (forerunners of today's 89th Airlift Wing), the C-87A carried senior military and government officials on numerous trips. Apparently, safety concerns for the first lady were less stringent, for the C-87A carried Eleanor Roosevelt on a March 1944 tour of American military installations in the Caribbean and Latin America.

Guess Where II continued to haul dignitaries until its final flight on October 30, 1945, when its destination was Walnut Ridge, Arkansas. Here, the first plane designed for a president—which never carried a president—was ignominiously retired and later scrapped.

TOP: This C-87A (serial number 41-24159) had a side cargo door equipped with a ramp for President Roosevelt's wheelchair. The transport is seen during a tour of 14th Air Force bases in China in August 1943. From left, the uniformed men are Maj. Gen. Claire L. Chennault, New York Senator J. M. Mead, Massachusetts Senator Henry Cabot Lodge Jr., Kentucky Senator A. B. Chandler, an unidentified Chinese officer, Lt. Gen. Joseph W. Stillwell, Georgia Senator R. R. Russell, Maine Senator Ralph O. Brewster, an unidentified man, Brig. Gen. Edgar E. Glenn, and another unidentified officer. *US Army Air Forces via Alan Griffith*

BOTTOM: The Individual Aircraft Record Card for C-87A (serial number 41-24159) was delivered on June 2, 1943, and served its entire career with the Air Transport Command. The Liberator departed for the Eighth Air Force in England (codenamed "UGLY") on July 29, 1943. *USAAF via Nicholas A. Veronico*

Continued from page 31

Roosevelt traveled to conferences in Tehran and Cairo in November 1943 aboard another TWA C-54 Skymaster, rather than using the C-87A Liberator Express *Guess Where II* that had been built for his use. The C-87A never carried a president, and Roosevelt's own personal C-54 did not become available until eight months after the November 1943 meetings.

This record of presidential travel has often proven confusing to those who try to write about it today. It is important to remember that Roosevelt was blazing a trail. Once he established a precedent by demonstrating that the occupant of the White House could travel abroad and fly in an aircraft, it was never seriously questioned again. A few recalcitrant taxpayers undoubtedly still believe, even in the twenty-first century, that their dollars would be better spent keeping the president at work in his office. Some certainly would argue that he could travel a little less ostentatiously than he does today. But after Roosevelt, aerial travel by presidents became the norm.

A NEW AIRCRAFT

On June 12, 1944, pilot Maj. Henry T. "Hank" Myers traveled to the Douglas Aircraft plant in Santa Monica, California. Myers's mission was to pick up a new C-54 Skymaster transport and deliver it to Washington National Airport, where it would be assigned to the 503rd Army Air Base Unit of the Air Transport Command. The unit was a precursor of today's 89th Airlift Wing, Air Mobility Command, and was part of the Army Air Forces, the predecessor to today's Air Force.

The C-54, to all appearances, was a standard, four-engine transport designed for long-range flying, one of a new generation of planes able to span vast stretches of ocean with hardly any refueling stops. But Myers's C-54 was special. It was the first transport designed and built to be a presidential airplane. The term "Air Force One" did not yet exist, and C-54C Army Air Forces serial number 42-107451 was known at first simply as "Project 51" and later as the "Flying White House." Most in Washington referred to

During President Roosevelt's February 1945 trip to the Yalta Conference in Crimea, Russia, the serial number of Douglas VC-54C 42-107451 *Sacred Cow* was changed to 42-72252 in an attempt to confuse enemy agents. *Courtesy National Museum of the US Air Force*

The elevator built into the fuselage of the *Sacred Cow* is displayed alongside the aircraft at the National Museum of the United States Air Force in Dayton, Ohio. The elevator enabled a wheelchair to be lifted from ground level up to the main deck of the C-54. *Courtesy National Museum of the US Air Force*

Franklin D. Roosevelt's personal plane by the informal name it had been given: the *Sacred Cow*.

Retired US Air Force Maj. Robert C. Mikesh, an authority on presidential aircraft, describes the way the *Sacred Cow* had been tailored to Roosevelt's needs: "[The] most unique feature of this aircraft was that it contained a battery-operated elevator, located aft of the main passenger cabin, which could lift a passenger directly from the ground to the cabin-level floor. This elevator represented a security measure as well as a convenience. The president's need to walk with crutches or use a wheelchair was an important factor in developing 'Project 51.' In the past, it had been necessary to construct bulky ramps to aid Mr. Roosevelt in embarking or debarking from an airplane. The very presence of such ramps at a foreign airfield suggested the impending arrival of FDR. Advance notification of the president as an incoming passenger was, of course, undesirable during World War II. The elevator eliminated the need for telltale ramps."

The aircraft had a bulletproof picture window, and a 7.5-by-12-foot stateroom occupied the aft portion of the cabin, providing seating for seven. Included was a sofa that opened electrically into a bed, two electrically folding chairs, and a full galley. In addition to pilot Myers, a crew of six manned the airplane. Myers's copilot was Capt. Elmer Smith, who later retired as a colonel. The stewards (who would be called flight attendants in today's air force jargon) were nicknamed "hotcuppers" because the plane's galley was equipped with electric hot cup devices to heat coffee and soup—"not very luxurious by today's standards," according to former flight attendant John Haigh.

The interior of Roosevelt's C-54 was furnished with upholstery of blue wool. Draperies at the windows were made of blue gabardine, on which was embroidered the insignia of the army, navy, marines, and coast guard. The aircraft was provided with its own set of dishes, silverware, and other amenities.

The *Sacred Cow* had been built for presidential use in part because the Secret Service felt the German U-boat threat made it unsafe for the chief executive to travel by sea across the Atlantic. Roosevelt, following his earlier trips on a Boeing 314 Clipper and a stock C-54, used the *Sacred Cow* for only one overseas trip, but it was his most important.

TRAVEL ON THE *SACRED COW*

Crossing the Atlantic aboard the heavy cruiser *Quincy*, the president arrived on the island of Malta. On February 3, 1945, he boarded the *Sacred Cow* for a trip to Yalta in the Soviet Union. He met with Britain's Winston Churchill and Russia's Josef Stalin until February 12. It was the first summit conference of the three leaders to discuss Allied strategy. Roosevelt then reboarded the *Sacred Cow* for a flight to Cairo, Egypt, where he rejoined the *Quincy* for the voyage home.

When Roosevelt died on April 12, 1945, the *Sacred Cow* became Harry S. Truman's aircraft. As the thirty-third president (1945–1953), Truman became the first chief executive to fly on a regular basis. He used the *Sacred Cow* to travel to Kansas City on his way to visit his hometown of Independence, Missouri. This was the first domestic air trip by a president while in office.

The *Sacred Cow* performed White House duties until 1947. Even after it was no longer a flying White House, the *Sacred Cow* soldiered on. Retired Col. Ancil Baker was one of many who flew the C-54 extensively from 1948 to 1963, after it was no longer employed for presidential travel.

Colonel Baker recalls, "The *Sacred Cow* was turned over to Headquarters Command, US Air Force at Bolling Air Force Base in Washington, DC. I flew it quite a bit. I took it to Mexico City, Ottawa and Goose Bay (Canada), Iceland, Wiesbaden (Germany), Copenhagen (Denmark), Paris, Madrid and Lajes (Spain), and Puerto Rico. We carried the air force band around.

"Our crew consisted of pilot, copilot, and flight engineer. We had no radio operator and no steward. The aircraft still had an elevator, a private room for the president, an overstuffed chair with an oversized window, and a desk. On the wall facing the president's desk was a huge seascape—Roosevelt having been a naval person all his life. There was a bed on one side of the office and a passageway on the right. The aircraft still had Roosevelt's elevator. We used the elevator to store cheap foreign liquor that we brought home from overseas."

Later, Baker saw the *Sacred Cow* again "in bits and pieces" at the National Museum of the United States Air Force in Dayton, Ohio, where it is displayed intact today. "They were replacing the Douglas [Aircraft Company] emblem on the control yoke, so they gave me the original."

In 1947, the *Sacred Cow* was replaced for presidential travel by a C-118 named the *Independence*, also piloted by Col. Henry T. "Hank" Myers. On December 4, 1961, Maj. Brooke E. Allen, commander of the air force's Headquarters Command in Washington, carried out the transfer of the *Sacred Cow* to the National Air Museum, predecessor of today's National Air and Space Museum. Subsequently, the museum released the aircraft for further transfer to the air force museum at Wright-Patterson Air Force Base in Dayton, Ohio. Museum experts studying the plane have discovered that the *Sacred Cow* had a unique aileron configuration not found on other C-54s; as displayed today, the plane has a standard C-54 outer wing, but all other parts are original.

INTRODUCING THE *INDEPENDENCE*

In 1946, the Army Air Forces made arrangements with Douglas Aircraft to acquire a production DC-6 airliner for executive use by the president.

The prototype for the DC-6 series began flight tests at the Santa Monica, California, factory on February 15, 1946. At this time, the DC-6 was the standard against which other airliners were measured. It was powered by four 2,100-horsepower Pratt & Whitney R-2800-34 Double Wasp CA15 engines, replacing the R-2000 Twin Wasp used on the DC-4. This was a proven engine with tens of thousands of hours of wartime experience in planes such as the P-47 Thunderbolt and F4U Corsair. It was an eighteen-cylinder, twin-row radial, air-cooled engine with a single-stage,

RIGHT: A rare photo of President Harry S. Truman's VC-118 *Independence* shows the stylized eagle's beak painted yellow. It was determined that the pigment in the yellow paint interfered with the operation of the nose-mounted weather radar, and it was subsequently painted white. *Courtesy National Museum of the US Air Force*

BOTTOM RIGHT: Douglas VC-118 46-0605 was named *Independence* in honor of President Truman's hometown of Independence, Missouri. Note the large three-pane picture window in the aft fuselage. *Courtesy National Museum of the US Air Force*

two-speed integral supercharger. The version of the DC-6 that became President Truman's *Independence* also became the first US military transport to use and test reversible-pitch propellers, as well as the first with water injection in the engine for added thrust on takeoff. The presidential airplane had greater fuel capacity than the standard DC-6, enough for an absolute range of 4,400 miles.

The *Independence* was equipped with an experimental weather radar (the first in a DC-6), a radar altimeter, autopilot, and other advanced navigation equipment. With a cruising speed of 300 miles per hour and a normal range under standard operating conditions of 3,000 miles, the *Independence* could reach any location in the continental United States nonstop.

OPPOSITE: President Truman takes time to shake the hand of each member of his police escort before boarding *Independence*. *William T. Larkins collection*

BELOW: Ascending the airstairs provided by Western Airlines, President Truman boards *Independence*. Note the presidential seal on the door's interior. *William T. Larkins collection*

American Airlines relinquished the twenty-ninth position for a DC-6 on the production line to enable the army to acquire an early aircraft, constructor's number 42881, and the army assigned it the serial number 46-505 and the designation C-118. The plane became the only C-118 ever to serve in the air force, although the service later acquired numerous DC-6A airplanes, designated C-118A.

The new plane for President Truman differed from civilian DC-6s because it had three closely grouped windows on the rear starboard side of the fuselage. This was the location of the presidential stateroom, entered through a bleached-mahogany door bearing the Great Seal of the United States. The room was decorated in chocolate brown, dark blue, and gray. The interior of the aircraft was configured to carry twenty-five passengers (compared with fifty-two in the airliner version). The main cabin could sleep twelve passengers.

Ever sensitive to perceptions, officials in Washington grimaced when an aviation magazine began calling the new plane the *Sacred Cow II*. Throughout the history of presidential air travel, officials have worried about the American taxpayer perceiving the chief executive's airplane as too glamorous or, worse, too ostentatious. To head off the unwanted name, Colonel Myers chose the name *Independence*, which was the name of Truman's hometown in Missouri but also, in Myers's view, a name with a "national flavor."

On July 27, 1947, Truman was aboard the *Independence* when he affixed his signature to the National Security Act, which among other things established the US Air Force as a separate

ABOVE: After serving President Truman, Douglas VC-1118 (serial number 56-0505) was returned to the air force's inventory to serve as a VIP transport. The aircraft is seen at Wright-Patterson Air Force Base, Ohio, on August 15, 1970, after its retirement to the air force museum's collection. *Col. J. Morris, USAF ret., via J. G. Handelman*

OPPOSITE: The view of *Independence's* cockpit as it sits today, preserved at the National Museum of the United States Air Force in Dayton, Ohio. *Courtesy National Museum of the US Air Force*

RIGHT: The radio operator's station on the flight deck of *Independence*. *Courtesy National Museum of the US Air Force*

military service branch, having equal footing with the army, navy, coast guard, and marines. The act also created the Department of Defense and the Central Intelligence Agency, but to those who served in the nation's military flying arm, the creation of an independent air force was much welcomed, if long overdue.

On January 6, 1948, Col. Francis T. Williams replaced Myers as the presidential pilot. Myers returned to his civilian job as a pilot with American Airlines. The two men began a long tradition of presidential pilots that continues today, in an era when the job has become more and more political. Since the 1980s, the authority and influence of the presidential pilot has varied according to the individual and the administration. What began as a prestigious but straightforward military assignment would eventually evolve into an assignment away from the mainstream of the military and close to the staff of the White House.

TOP: The office area of *Independence*, with the conference table, pull-down maps, divan, and large picture window at left. *Courtesy National Museum of the US Air Force*

BOTTOM: The galley of President Truman's *Independence* has a four-burner cooktop, oven, and nearly full-size refrigerator. *Courtesy National Museum of the US Air Force*

Independence had a crew of nine and was configured to carry twenty-five passengers. The VC-118 was retired from air force service in 1965 and was restored to its presidential eagle paint scheme from 1977 to 1978 by staff at the National Museum of the United States Air Force. *US Air Force photo by Ken LaRock*

On June 1, 1948, a new Military Air Transport Service (MATS) replaced the former Air Transport Command. The unit operating presidential airplanes at Washington National Airport was established as the 1254th Air Transport Squadron on October 1, 1948, following the consolidation of several units. The presidential outfit would undergo several name changes before evolving into today's 89th Airlift Wing; at the time of its founding, it was commanded by Williams.

During the years when Truman continued to be an occasional flyer onboard *Independence*, another change of military nomenclature took place. On August 1, 1952, the 1254th was redesignated a group, and two squadrons were established. The 1298th Air Transport Squadron was assigned responsibility for overseas missions with four-engine aircraft from Washington National Airport. The 1299th Air Transport Squadron, soon to be equipped with twin-engine Convair transports, was located across the Potomac River at Bolling Air Force Base, DC, for domestic missions carrying the president and other VIPs.

"Truman was not that much of an air traveler," remembers one officer who has followed the development of presidential travel. "Remember, during his era the railroad train was still the primary way of getting around. But he did make use of the *Independence* and he was respected and liked by those who flew him."

The Jet Age

After World War II, Supreme Allied Commander in Europe Dwight D. Eisenhower traveled in a C-121A Constellation named *Columbine* (in honor of the state flower of Colorado, the home state of Eisenhower's wife, Mamie). After moving into the White House on January 20, 1953, Eisenhower selected as his personal aircraft another C-121A Constellation operated by the 1254th Air Transport Group at Washington National Airport. The plane was quickly dubbed *Columbine II*. The name was painted in flowery script across the nose of the aircraft above a likeness of a blue columbine. The presidential pilot, air force Col. William G. "Bill" Draper, flew the aircraft, led the crew, and served Eisenhower as a kind of extra military advisor. Draper had also been Ike's pilot in Europe.

A little-known fact about Eisenhower is that he was a pilot himself, indeed the first president licensed to pilot an airplane, having soloed a Stearman PT-13 biplane trainer in the Philippines in 1936. Eisenhower was issued a private pilot's license on July 5, 1939, by the Commonwealth of the Philippines. He also had a Certificate of Competency from the US Civil Aeronautics Authority. Eisenhower logged 350 hours of flying time from July 1936 to November 1939. Aboard his new presidential aircraft, however, Eisenhower was strictly a passenger.

By the time *Columbine II* was tapped for presidential duty, the aircraft had already carried President-elect Eisenhower on his famous Far East trip of November 1952, fulfilling his campaign pledge to visit the war in Korea.

OPPOSITE: Soldiers stand guard around President Eisenhower's Lockheed VC-121 (serial number 48-0610) *Columbine II* as it sits at Washington National Airport. This is the only major presidential aircraft in public ownership and is being restored to fly by Karl D. Stoltzfus Sr., of Dynamic Aviation in Bridgewater, Virginia. *National Park Service*

COLUMBINE II

Ike's C-121A was essentially a Lockheed L-749 airliner with strengthened floors and a port rear fuselage cargo door. A graceful and elegant aircraft with its sharklike fuselage and triple tail, *Columbine II* was powered by four 2,500-horsepower Wright R-3350 (749C-18BD1) Cyclone piston engines.

To become a flying White House for the new president, the Constellation was removed from service and modified with the installation of a customized suite amidships. The 20-foot-long interior cabin was equipped with two brown leather swiveling chairs, a table, and two davenport couches that opened into beds. A large lavatory was installed in the tail section. Forward of the presidential cabin were two duplicate cabins, each providing seating for sixteen or sleeping berths for eight.

Because of its very important passenger, *Columbine II* became a VC-121A, the *V* prefix signifying its role carrying dignitaries. The aircraft was equipped with the latest in flying

ABOVE: First Lady Mamie Eisenhower christens Lockheed VC-121E (serial number 53-7885) *Columbine III* at Washington National Airport on November 24, 1954. *Courtesy National Museum of the US Air Force*

LEFT: Future president Dwight D. Eisenhower boards an open-top Lincoln Continental for a motorcade from the airport to the 1952 Western Governor's Conference in Boise, Idaho. Eisenhower made extensive use of the Lockheed Constellations as Air Force One during his term. *William T. Larkins collection*

President Dwight D. Eisenhower and First Lady Mamie Eisenhower exit *Columbine II* during the chief executive's first year in office. The president and first lady walked through the crowd of people to their limousine, which is out of sight of the photographer. *US Air Force*

instrumentation, including weather radar and long-range navigation (LORAN) equipment. A crew of eleven operated the VC-121A.

It should be noted that almost every aircraft operated by the air force to support the president wore the *V* prefix at one time or another. The prefix has been added or removed under different administrations.

Quite remarkably, *Columbine II* is under restoration to fly and is the only presidential aircraft to have ended up in the hands of a civilian owner. A presidential seal from the aircraft, bearing forty-nine stars (as the seal did before Hawaii was admitted as a state), is on display at the US Army Quartermaster Museum at Fort Lee, Virginia.

COLUMBINE III

In 1954, Air Force leaders decided it was time for President Eisenhower to have a new aircraft for his personal transportation. Ike's C-121A Constellation, the *Constellation II*, was racking up a high number of flight hours (it had accumulated four thousand before becoming the president's mount) and no longer represented up-to-date technology. With considerable input from pilot Draper, officials decided

on a Lockheed L-1049C Super Constellation, a longer, heavier aircraft with greater fuel capacity and more interior space.

The new aircraft was known in military jargon as a VC-121E Super Constellation (serial number 53-7885, constructor's number 4151) and was given the inevitable name *Columbine III*. It had been originally procured by the navy as an R7V-1 model, Buno. 131650, but was transferred to the air force before leaving the Burbank, California, factory.

The air force accepted Ike's VC-121E at the factory on August 31, 1954. Draper brought the aircraft to Washington National Airport on September 10. First Lady Mamie Eisenhower conducted a christening ceremony on November 24 using a bottle of Rocky Mountain spring water flown in from her beloved Colorado. That day, President Eisenhower made his first trip in the new aircraft, flying to Augusta, Georgia.

Columbine III ultimately spent six years as a presidential aircraft. Unlike today's Air Force One, it was not used exclusively by the president. When Ike wasn't traveling, the VC-121E carried other government officials. This was probably a good deal for the taxpayers. After all, Ike only flew about 30,000 miles per year, a figure that today would rank him in the lowest category of an airline's frequent flyer club.

LEFT: VC-121E (serial number 53-7885) *Columbine III* is serviced while President Eisenhower gives a speech at a nearby civic function. This aircraft began transporting the president in 1954 and served until Eisenhower left office in 1961. The VC-121E had a 4,000-mile range, nearly double the range of the VC-121A that it replaced. *Courtesy National Museum of the US Air Force*

BELOW: The cockpit of VC-121E *Columbine III* as it sits today, preserved at the National Museum of the United States Air Force in Dayton, Ohio. *Courtesy National Museum of the US Air Force*

Beginning in March 1957, the US Air Force operated a pair of Bell UH-13J Sioux helicopters for presidential transport duties. UH-13J (serial number 57-2728) is seen undergoing a series of flights to and from the South Lawn of the White House on June 14, 1957, to familiarize air force pilots Maj. Joseph E. Barrett and Capt. Laurence R. Cummings with the flight path and landing marks. President Eisenhower became the first sitting president to make a helicopter flight when he flew from the White House lawn. The US Army and Marine Corps took over presidential helicopter transport duties in 1958, and it became the exclusive domain of the US Marine Corps in 1976. *Courtesy National Museum of the US Air Force*

Ike's VC-121E was powered by four Wright R-3350-34 (civilian model 972TV18DA-1) Turbo Compound engines. The new presidential transport was 18 feet longer than its predecessor, offered a range of 3,500 miles, and was able to fly at a speed of 355 miles per hour. Long after Eisenhower became the first president to have a jet aircraft assigned to him, the VC-121E continued to fly transport missions with the air force. Late in its career, it was redesignated C-121G.

INTRODUCING THE BOEING 707

During the Eisenhower administration, the jet engine became the primary source of power for new military, civil, and commercial aircraft, replacing propellers. A vigorous US aircraft industry recognized the potential to make billions of dollars by converting the world's airlines from props to jets. A lucrative

sidelight would be converting the air force's fleet of air-refueling tankers from propeller-driven to jet-powered aircraft.

In an era when corporations took more risks than they do today, Boeing gambled $16 million on the revolution it saw coming. On July 15, 1954, a yellow and maroon four-jet prototype aircraft took to the skies over Seattle. It was dubbed the Boeing 367-80, taking its model number from the Stratocruiser series of prop-driven transports. To those who worked on it and flew it, the aircraft was called simply the "Dash 80." Until the plane was made public, its misleading name veiled the corporate secret that the aircraft was no Stratocruiser and had no props.

It helped Boeing that Gen. Curtis E. LeMay of the Strategic Air Command wanted the new aircraft as an air-refueling tanker. The United States was fielding jet bombers capable of cruising speeds as high as 500 miles per hour, including the B-47 Stratojet, B-52 Stratofortress, and the B-58 Hustler. They needed a gas station in the sky that could keep up.

On August 5, 1954, just three weeks after the maiden flight of the Dash 80, the USAF announced that it would purchase a limited number of jet tankers. The service eventually acquired 820 Boeing KC-135 Stratotankers, based on the Dash 80 and known in manufacturer's jargon as Boeing 717s. Test pilots "Dix" Loesch and "Tex" Johnston took the first KC-135A aloft for its maiden flight on August 31, 1956.

The first KC-135A became operational on June 28, 1957. By then, the race was on to field jet airliners, and the air force was seriously looking to the jet as a possible replacement for Eisenhower's *Columbine III*. By then, the Dash 80 test ship was equipped as a civil demonstrator for airlines. Having given birth to a tanker, it would now spawn an airliner.

The first production Boeing 707 (constructor's number 17586) took to the air at Renton, Washington, on December 20, 1957. This plane wore the civil registration N708PA, making it appear to be the second of an initial batch of twenty 707s for Pan American World Airways. The second aircraft in the series had been chosen for the special registration N707PA. As late as the second half of the 1950s, some skeptics were saying that these jetliners would never make the grade. In Britain, the Bristol company was still trying to convince the world that its Britannia airliner—pulled through the sky by propellers—

continued on page 52

EISENHOWER'S L-26B AERO COMMANDER

Today, Americans are accustomed to seeing the president travel in a giant aircraft, familiar to all as Air Force One. But in the 1950s, some air force members flew the president in a much smaller plane. For short trips from Washington to his beloved farm in Gettysburg, Pennsylvania, which had a grass runway, Dwight D. Eisenhower favored the air force's twin-engine L-26B Aero Commander.

In 1955, the air force issued a contract to Aero Design and Engineering Company of Bethany, Oklahoma, for a military version of the Aero Commander business aircraft. The L-26B (its *L* designation signifying a liaison function) became the first presidential aircraft to carry the now-familiar blue-and-white paint scheme. In the 1950s, the color scheme was known as Baltic Blue and Polar White.

On June 3, 1955, the first flight of a president in a light plane took place when Ike made the thirty-two-minute trip to Gettysburg in an Aero Commander 560 loaned to the White House by the manufacturer while the air force L-26Bs were still being built.

The air force eventually ordered fifteen L-26B airplanes, all based on the improved Aero Commander 560A. They were powered by two GO-480-G1B6 Lycoming piston engines providing 295 horsepower. The L-26B carried a crew of just

President Eisenhower and presidential pilot Col. William G. Draper stand with the Aero Commander that often served as Air Force One from 1956 to 1960. *National Museum of the US Air Force*

two pilots, plus four to eight passengers at a cruising speed of 250 miles per hour.

The planes were stationed at Bolling Air Force Base in Washington, DC. The presidential unit was the 1299th Air Transport Squadron of the 1254th Air Transport Wing, a predecessor of today's 99th Airlift Squadron, 89th Airlift Wing, at Joint Andrews Base, Maryland.

Ike used one of these planes for frequent trips to Gettysburg. Fitted temporarily with a bed, the planes shuttled him back and forth during his recovery from a heart attack in the summer of 1955. He later flew in an L-26B to a March 26, 1956, meeting with US and international dignitaries at White Sulphur Springs, West Virginia.

The last two of the fifteen aircraft in the air force's purchase were modified on the production line to Aero Commander 680 status. These two planes had improved Lycoming GSO-480-A1A6 engines providing 340 horsepower each, carried additional fuel, and were about 20 miles per hour faster.

The improved Commanders were called L-26Cs. Eisenhower made his first flight in an L-26C on June 6, 1956. One of these aircraft carried West German Chancellor Konrad Adenauer from New York City to Gettysburg. Eisenhower and Adenauer then traveled together in the aircraft to Washington.

The air force's Aero Commanders lost their presidential duties when Ike left office on January 20, 1961. By then, the radio call sign for Air Force One had been adopted to refer to any air force plane carrying the chief executive. Also by that time, helicopters were able to take over short-range flights of the kind for which the L-26B and L-26C had been purchased.

The L-26B and L-26C fleet performed many missions in addition to flying the commander-in-chief. They were considered extremely reliable for general utility missions and could land at small airfields where larger transports could not gain access.

In a change of military nomenclature that took place in 1962, the L-26B became the U-4A (for utility), and the L-26C became the U-4B. At least one U-4B was used in the 1960s by the Air Force Academy for cadet parachute training and for the academy's skydiving team.

The L-26B (or U-4B) used by Eisenhower is now on display at the National Museum of the United States Air Force in Dayton, Ohio.

Aero Commander U-4A (serial number 55-4634) on the ramp at Andrews Air Force Base, Maryland, on April 2, 1966. This aircraft, and the other U-4As in the background, continued to serve the executive transport needs of the US Air Force. The air force purchased a total of fourteen U-4As. *J. G. Handelman*

continued from page 49

was the wave of the future. Bristol engineers asserted that the Boeing 707 design was too revolutionary to be harnessed for practical work, that the airframe and wing would require six engines rather than four, and that the Boeing's aircraft was a monster, impossibly big, heavy, capacious, and expensive, needing long runways and ready to bankrupt airlines. Boeing had risked more than the company's net worth to build the prototype 367-80, was heavily in hock for the earliest 707s, and seemed to be taking a desperate gamble, but the airlines liked what Boeing was offering and were ready to leap into the future.

In an era when fewer than 10 percent of Americans had even seen the interior of any aircraft, airline travel was still the province of the privileged, and it took place in a realm where the future was now. Promoting glamour and sophistication, the airlines went for the graceful and futuristic Boeing 707 without hesitation and, surprisingly, with relatively few teething troubles.

The 707 made its first airline revenue flight on October 26, 1958, a scheduled service from New York to Paris. The Douglas DC-8 followed suit on September 18, 1959, and the Convair 880 on May 14, 1960.

Thinking much further ahead than Bristol or anyone else who advocated propellers, the air force picked the Boeing 707 for presidential duty. In military parlance, the aircraft would be called the C-137A or VC-137A, dropping and regaining the *V* prefix several times in its career.

April 7, 1959, marked the first flight of the first military Boeing 707-153, a VC-137A model (58-6970, later to be named *Queenie*). Soon afterward, Boeing delivered *Queenie* to the 89th Airlift Wing at Andrews Air Force Base, Maryland. This aircraft was not earmarked for presidential travel but would become the first jet to carry a president.

Unlike the 707 airliner, the military VC-137A introduced a special communications section located in the forward fuselage, ahead of an eight-seat passenger compartment. The center cabin was configured as an airborne headquarters with a conference table, swivel chairs, a film projector, and divans convertible to beds. The aft compartment contained fourteen reclining passenger seats. In all, the VC-137A would have been able to carry forty passengers on a typical VIP trip.

TOP: President John F. Kennedy deplanes from VC-137B (serial number 58-6970), nicknamed *Queenie*, at Naval Air Station Alameda, California, on March 23, 1962. The president flew into the San Francisco Bay Area to visit the Lawrence Berkeley Radiation Laboratory and to speak at the Charter Anniversary Ceremony at the University of California Berkeley. *Robert Knudsen. White House Photographs. John F. Kennedy Presidential Library and Museum, Boston*

BOTTOM: Upon arriving at Los Angeles International Airport in California, President John F. Kennedy walks with Attorney General Stanley Mosk of California (facing away from the camera) and Governor Edmund G. "Pat" Brown (right). VC-137B (serial number 58-6970) served as Air Force One for the November 18, 1961, trip to the West Coast. *Robert Knudsen. White House Photographs. John F. Kennedy Presidential Library and Museum, Boston*

OPPOSITE: Full fuselage view of 58-6970's paint scheme while parked at the St. Louis headquarters of McDonnell Aircraft Co. McDonnell Aircraft built the Mercury space capsules and President Kennedy was on hand to tour the facility and greet employees. *McDonnell Aircraft Corp., Courtesy of Harry S. Truman Library*

On August 26, 1959, Eisenhower became the first president to fly aboard a jet, on *Queenie*, which was eventually used by seven presidents even though it was always considered only a backup. He flew to West Germany for a meeting with Chancellor Konrad Adenauer. At this juncture, *Queenie* was attired in her initial paint scheme, which included fluorescent red trim on nose, wingtips, and tail, a precaution taken as a result of the midair, broad-daylight collision of a United Airlines Douglas DC-7 and a Trans World Airlines Lockheed Constellation over the Grand Canyon in June 1956.

Andrews's elite air wing took delivery of three VC-137As in this series (serial numbers 58-6970-6972, constructor's numbers 17925-17927), the first of which was *Queenie*, but none of which were purchased with the intent of transporting the president. Only the first ship served repeatedly and frequently as a backup to presidential aircraft. Possibly without Eisenhower's knowledge, the CIA outfitted *Queenie* with secret reconnaissance cameras in preparation for Ike's planned 1960 summit meeting in Moscow. That meeting was scuttled, ironically, because of the shooting down of Francis Gary Powers in his U-2 spy plane over the USSR on May 1, 1960.

Almost forgotten, however, is the fact that Eisenhower did not travel to France for the proposed summit by jet, meaning via *Queenie*. He was in Paris from May 15 to May 19, 1960. The term Air Force One had been in use for several years by that date. In 1962, *Queenie* carried astronaut John Glenn to Washington the day after his orbital flight.

Known initially as the VC-137A, this aircraft was powered by early, 13,500-pound-thrust Pratt & Whitney JT3C turbojet engines, with water injection and distinctive engine pods having multipipe nozzles. The large noise-suppression nozzle with twenty separate tubes routinely became covered with soot from wet takeoffs, when black smoke was emitted in impressive amounts. Not long after the planes' delivery, the air force retrofitted its trio of VC-137As with 18,000-pound-thrust Pratt & Whitney JT3D-3 turbofan engines and redesignated them as VC-137Bs.

A NEW CALL SIGN

During the mid-1950s, the president's aircraft received a new radio call sign. There is disagreement, however, as to exactly when and how it happened—even among those who served

LEFT: President John F. Kennedy boards Air Force One for his flight to Hyannis Port, Massachusetts, following his birthday dinner on May 27, 1961, at the National Armory hosted by the Democratic National Committee. After the dinner, Kennedy boarded a VC-118A at the Military Air Transport Service (MATS) terminal at Washington National Airport. *Abbie Rowe. White House Photographs. John F. Kennedy Presidential Library and Museum, Boston*

RIGHT: President John F. Kennedy boards a VC-118A, most likely serial number 53-3240, serving as Air Force One after attending a political rally in support of Richard J. Hughes, Democratic candidate for governor of New Jersey. From left, New Jersey Governor Robert B. Meyner, Richard J. Hughes (shaking hands with the president), and President Kennedy. The propeller-driven VC-118 was used as Air Force One for the short flight from Andrews Air Force Base, Maryland, to Mercer County Airport in Trenton, New Jersey. *Cecil Stoughton. White House Photographs. John F. Kennedy Presidential Library and Museum, Boston*

Presidential transport VC-118A (serial number 53-3240) departs Andrews Air Force Base, Maryland, on February 19, 1966. During the Kennedy administration, this aircraft served extensively as Air Force One, taking the president into a number of smaller airports, such as Hyannis Port, Massachusetts, where the jet-powered VC-137Bs could not land. *R. C. Sullivan via J. G. Handelman*

as crew members on presidential aircraft. One former radio operator aboard *Columbine II* reports that the name Air Force One was created in a group conference at Washington National Airport. Others, however, insist that confusion during a flight was responsible for the new term and that Ike's pilot, Draper, suggested it.

As the story goes, "Air Force 610" (the designation for *Columbine II*) was on a flight to Florida and was receiving radio instructions when "Eastern 610" (a commercial flight by Eastern Airlines) came on the airwaves. Whatever confusion took place was brief. No mishap occurred, but Draper and others saw the event as a wake-up call. The president already enjoyed top priority when traveling the nation's airways (a "priority 1" according to some), but to prevent accidents, the chief executive needed to be quickly and readily identified.

The term Air Force One was born.

From about 1956 onward, any air force aircraft carrying the president became Air Force One. The aircraft of other services transporting the chief executive used call signs accordingly: Army One, Marine One, and so on. It would be more than forty years in the future before a sitting president was flown in a navy aircraft. The coast guard, always a little different, continues to use the call sign Coast Guard One for its own boss, the secretary of transportation. Should the president fly on a civilian aircraft, it would use the call sign Executive One.

Today, everyone recognizes the term Air Force One. The term has become the title of a blockbuster movie starring Harrison Ford, a television special by *National Geographic* magazine, and a documentary by the Discovery Channel. It is familiar to air traffic controllers from Moscow to Monrovia. To many, Air Force One is a symbol of sovereignty and influence.

RUDE AWAKENING

Once the new call sign for the president's aircraft had been coined, it was not widely announced or publicized. Even in the air force, many did not know about it right away. An anecdote clarifies how one officer learned.

In May 1960, air force Capt. John W. Darr was serving with the 99th Bombardment Wing, a B-52 Stratofortress unit at Westover Air Force Base, Massachusetts. He recalls, "I was one of two men who ran what we called the Current Operations Section. Because we were one of the very early B-52 outfits (the third, in fact), we were frequently tasked with performing 'higher headquarters directed' missions. I was also the point of contact within the wing for all matters related to coordination with federal aviation authorities.

"When we'd have a special mission planned, I'd coordinate the plan with Boston Center, who in turn would pass it on to the Central Altitude Reservation Facility (CARF, which was located in Kansas City.

"Every such mission was assigned a priority. As I recall, priority 2 was for high-ranking individuals, 3 was an emergency in progress or in-flight refueling, 4 was for missions of some importance, while 5 was for routine military activity. I don't recall that I was aware of a priority 1.

"When flying on a priority 4 or 5, it was common for me to receive a phone call—often in the middle of the night, no less—from some guy at CARF. Over the phone we'd resolve conflicts, and life would proceed.

"One day, our outfit was in the midst of keeping two B-52s airborne around the clock on airborne alert. Because they were carrying nuclear weapons, we were operating on a priority 2. As a logical consequence, those midnight phone calls did not occur. If there were conflicts between our aircraft and others, I'm sure the other fellows were ordered to change their plans—but certainly not anyone with our top priority!

"You can imagine my surprise when I was awakened at 3 o'clock one morning by a call from CARF. The fellow on the other end of the line told me that I would have to alter the route, altitude, or timing of our two-ship B-52 cell somewhere out over the Atlantic, east of Boston. In my mind I was trying to determine what best to do when I finally realized that my priority 2 was being challenged. I asked him, 'Who's asking us to move?' I asked.

"He responded with, 'Air Force One.' I had never heard the term before. (Later, I found that no one else in my outfit had ever heard the term, either.) So I responded with, 'Who is that?'

"He retorted with, 'My God, man—-that's the PRESIDENT!'

"So I gave him a 'Really! OK, we'll move.' And I agreed to have our aircraft fly high enough to avoid the perceived conflict.

"The presidential mission that our B-52s had almost interfered with was the one carrying President Eisenhower to Paris for the Big Four summit, which the Soviet Union's Nikita Khrushchev refused to attend. My recollection is that Eisenhower was traveling in a Boeing 707 (a.k.a. *Queenie*)."

A NEW PRESIDENT

During Senator John F. Kennedy's 1960 presidential campaign, his primary mode of transportation was a Convair 240 named *Caroline*, after his daughter. His father purchased the aircraft for him. In November 1960, Kennedy won a narrow election to the presidency and soon afterward visited Eisenhower in the White House. According to one observer, Eisenhower believed Kennedy was "a bit of a political lightweight," so during their first meeting at the White House, Eisenhower picked up the phone and called for a "Marine Condition Green." Three minutes later, a helicopter landed on the lawn ready to evacuate the president. Eisenhower made this demonstration in order to show Kennedy the power of the presidency.

continued on page 60

RIGHT: VC-118A (serial number 53-3240) is today preserved at the Pima Air and Space Museum in Tucson, Arizona. *Rick Turner*

ARMY HELICOPTERS CARRY THE PRESIDENT

For eighteen years, between 1958 and 1976, the army shared in the job of providing air transport for the president of the United States.

The army's role is often forgotten today, as the public watches the president travel in Air Force One or the Marine Corps helicopter Marine One. But soldiers had a role in providing helicopter travel to five presidents, using a helicopter that became Army One when the president was on board.

In 1957, President Dwight D. Eisenhower began asking about the use of rotary-winged aviation as a means of presidential transportation. That year, Ike became the first commander-in-chief to fly aboard a helicopter when he took an air force Bell H-13J from the White House lawn to his retreat at Camp David, Maryland.

But although two H-13Js (known after 1962 as UH-13Js) were procured for presidential travel, the Secret Service questioned the margin of safety offered by

a single-pilot craft. Ike flew in the H-13J only once, and the air force never again operated a helicopter dedicated to the presidential transport mission.

Also in 1957, Marine Corps squadron HMX-1, already in existence for a decade as a test unit, became the helicopter squadron for the personal transportation needs of the president, but army leaders bristled. Ike was one of the most famous of army generals, after all, and army helicopter officers felt they could the job as well as the marines.

On January 1, 1958, the service activated the army Executive Flight Detachment at Davison Army Airfield, Fort Belvoir, Virginia. This squadron-sized unit, together with marine squadron HMX-1 at Quantico, Virginia, were to share the primary mission of the emergency evacuation of the president, his family, and other key government officials. In addition to this mission, these units furnished helicopter transportation for the president and others.

The army's aircraft of choice was the Sikorsky VH-34C Choctaw helicopter. This aircraft was almost identical to the HSS-1Z being used by the marine unit (which was redesignated VH-34D in 1962).

The VH-34 was powered by a 1,525-horsepower Wright R-1820-84 radial engine, was flown by two pilots, and could travel up to 120 miles per hour with a range of 210 miles, making it ideal for Ike's jaunts to Camp David and his farm in Gettysburg, Pennsylvania.

In a memo dated 1960, VH-34C pilot Lt. Col. William A. Howell noted that the VH-34C was "thoroughly reliable" for travel by the chief executive. The Secret Service agreed. On occasion, Eisenhower and his successor President John F. Kennedy were offered a choice. The president would arrive at Andrews Air Force Base, Maryland, step off his airplane, and find both army and marine VH-34s waiting for him.

The term Army One as a radio call sign for an army aircraft carrying the president came into use in 1960. Presidents Eisenhower, Kennedy, Johnson, Nixon, and Ford all flew in army helicopters. To iron out their differences in an early example of "jointness," the two services with rotary wing aircraft combined their administrative functions in an Army–Marine Corps Executive Flight Detachment, but competition remained a part of everyday life.

President Dwight D. Eisenhower emerges from Army One, a Sikorsky VC-34C, in 1958. *Courtesy Pima Air and Space Museum*

The VH-34Cs piloted by Howell and other army aviators with presidents as their cargo can still be viewed at museums. One is on display at the United States Army Aviation Museum at Fort Rucker, Alabama, another at the Pima Air & Space Museum in Tucson, Arizona.

The army VH-34Cs were eventually replaced by VH-1 Iroquois (Huey) and VH-3 helicopters during the Nixon years.

In 1976, in a cost-cutting measure prompted by President Jimmy Carter, the Pentagon decided to turn over the presidential helicopter mission to the marines, and it has been carried out by HMX-1 ever since. The army still provides support but no longer operates helicopters earmarked for the president.

RIGHT: *Dwight D. Eisenhower Presidential Library via Pima Air and Space Museum*

BELOW: *US Army Aviation Museum via Pima Air and Space Museum*

continued from page 56

The 1254th Air Transport Group was enlarged and redesignated the 1254th Air Transport Wing on December 1, 1960. The presidential wing was preparing to receive the first jet aircraft explicitly designed for presidential travel. This aircraft was the first of two special 707-353Bs ordered in 1961, designated VC-137C and assigned serial number 62-6000 (constructor's number 18461).

By then, the aircrews who transported the president were referring to themselves as the SAM Fox outfit. SAM was the abbreviation for Special Air Mission, and Fox was the old phonetic term for the letter *F*, as in flight, so the term was an acronym for "Special Air Mission Flight." The new presidential plane was Air Force One to the outside world, but to insiders it was SAM 26000.

SAM 26000 arrived at Andrews on October 10, 1962. When the spanking-new jet entered service, it was capable of traveling farther and faster than any other executive aircraft in the air force fleet. It could also operate from much shorter runways.

To all who maintained it, worked on it, and flew it, SAM 26000 was the most beautiful plane they had ever seen.

RIGHT: President and Mrs. Kennedy deplane from Air Force One at Love Field, Dallas, Texas, on the morning of November 22, 1963. The aircraft is VC-137C (serial number 62-6000), which played an important part in the events of this day. *Cecil Stoughton. White House Photographs. John F. Kennedy Presidential Library and Museum, Boston*

OPPOSITE: After President John F. Kennedy was shot to death in Dealey Plaza in downtown Dallas, Vice President Lyndon B. Johnson was quickly escorted to the safest location, one where the Secret Service could assure his safety—on board Air Force One at Love Field. With the casket carrying President Kennedy on board, Vice President Johnson takes the oath of the office of President of the United States with Jacqueline Kennedy looking on. From left to right: Assistant Press Secretary Malcolm Kilduff (lower left, holding dictating machine); a media liaison, Jack Valenti; Judge Sarah T. Hughes (administering oath); Representative Albert Thomas of Texas; President Johnson's secretary, Marie Fehmer (mostly hidden in back); Lady Bird Johnson; Dallas Police Chief Jesse Curry (face hidden by President Johnson's raised hand); President Johnson; President Kennedy's personal secretary, Evelyn Lincoln (mostly hidden); Representative Homer Thornberry of Texas (behind Lincoln); Jacqueline Kennedy; White House Secret Service agents Roy Kellerman and Thomas "Lem" Johns (both standing in doorway); Mrs. Kennedy's press secretary, Pamela Turnure (partially hidden in back); Representative Jack Brooks of Texas; unidentified (head down); Deputy Director of Public Affairs for the Peace Corps Bill Moyers (in back, mostly hidden behind Brooks); and President Kennedy's physician, Adm. Dr. George G. Burkley (on edge of frame). *Cecil Stoughton. White House Photographs. John F. Kennedy Presidential Library and Museum, Boston*

It represented a new age, just like the new and youthful president who seemed not to be living in the past but pointing to the future.

Encouraged by the president, First Lady Jacqueline Kennedy commissioned the noted designer Raymond Loewy to devise a new paint scheme for the White House's new jet. Loewy was, of course, designer of the Studebaker Avanti automobile, the Pennsylvania Railroad paint scheme, and the Ritz cracker logo. He also created the designs for the Coca-Cola bottle, the Lucky Strike cigarette package, and Greyhound buses.

Until now, the upper fuselage of an air force transport had borne the words "United States Air Force" or "Military Air Transport Service." Both were standard examples of military markings, but the Kennedys wanted to show that SAM 26000 represented more than just a part of the air force and, indeed, more than the entire air force.

Loewy may have been inspired by the blue-and-white paint design that had adorned Eisenhower's twin-engine L-26B Aero Commander. He cast aside standard military markings. Instead, SAM 26000 would bear the words "United States of America," making it an aerial ambassador of sorts. The presidential VC-137C wore an American flag on the tail

(with the union—the blue field containing 50 stars—facing forward on both sides of the fin, as correct usage dictated), a striking blue-and-white paint job, and large replicas of the presidential seal on both sides of the nose. To make the colors match as planned, Loewy even toned down the blue in the national insignia on the rear fuselage.

Except for the presidential seal, a similar paint scheme was eventually adopted by all of the VIP aircraft at Andrews.

Two days after arriving at Andrews, the aircraft made its first official flight, to Wheelus Air Base, Libya, to bring that country's crown prince to the United States for a visit. As the Cuban missile crisis loomed immediately after it entered service, SAM 26000 brought senators and congressmen from their home states to Washington since Congress was not in session at the time. President Kennedy flew 26000 for the first time in November 1962, when he and the first lady attended Eleanor Roosevelt's funeral in New York. In June 1963, Kennedy used the aircraft when he flew to Ireland and Germany, where he made his famous "Ich Bin Ein Berliner" speech. A month earlier, while taking a US delegation to Moscow, 26000 broke thirty speed records, including the fastest nonstop flight between the United States and the Soviet Union.

LEFT: President Kennedy's casket is transferred to Air Force One at Love Field in Dallas, Texas. Visible above the trailing edge of the wing are First Lady Jacqueline Kennedy; Special Assistants to President Kennedy Larry O'Brien, Kenneth P. O'Donnell, and Dave Powers; Assistant Press Secretary Malcolm Kilduff; Military Aide to President Kennedy Gen. Chester V. Clifton; secretary to Mrs. Kennedy, Mary Gallagher; and White House Secret Service agents Roy Kellerman, Dick Johnsen, Ernie Olsson, and Paul Landis. *Cecil Stoughton. White House Photographs. John F. Kennedy Presidential Library and Museum, Boston*

RIGHT: First Lady Jacqueline Kennedy and her secretary, Mary Gallagher, board Air Force One at Love Field in Dallas, Texas, following the transfer of President Kennedy's casket to the airplane. *Cecil Stoughton. White House Photographs. John F. Kennedy Presidential Library and Museum, Boston*

CARRYING KENNEDY

Kennedy frequently used SAM 26000 and its backup, *Queenie*. He also made regular use of the army and marine helicopters available to the chief executive. But Kennedy's favorite aircraft was a prop-driven C-118A, the military version of the Douglas DC-6 and a close relative of the *Independence*, used by Harry S. Truman.

This C-118A (serial 53-3240) had been delivered to the 1254th Air Transport Wing at Washington National Airport on December 23, 1955. Always a VIP transport, it apparently never transported Eisenhower. In July 1961, it moved with the wing to Andrews. During this period, it carried Kennedy to his home in Hyannis Port, Massachusetts, on numerous occasions. In the subsequent administration, it would be one of numerous aircraft to carry Lyndon B. Johnson to his Texas ranch. Air force records refer to this aircraft both as a C-118A and a VC-118A. Of all the aircraft in the history of presidential travel, this one is the most ambiguous. Was it intended from the beginning as a presidential transport, or was it another "almost" presidential aircraft that did, in fact, carry presidents on more than a few occasions? The record is unclear.

As for SAM 26000, it flew John and Jacqueline Kennedy on their visit to Dallas on November 22, 1963. When news that the president had been shot reached Love Field and 26000's commander, Col. James Swindal, the aircraft was prepared for immediate departure. Vice President Lyndon Johnson was also in Dallas that day. Fearing a wider conspiracy, Secret Service agents rushed Johnson to the safety of the air force aircraft. Because the communications equipment of 26000 was superior to that of the aircraft that Johnson had flown to Dallas, the decision was made that Johnson should wait aboard 26000 for Jacqueline Kennedy and her husband's body. Crew members felt it would be undignified for the former president's body to ride back to Andrews in the cargo hold; making room for the casket in the passenger compartment meant removing a partition and four seats from the rear of the aircraft. Before 26000 could leave Dallas, Johnson took the oath of office on board the aircraft. At Arlington National Cemetery, as the president's body was being lowered into the ground, 26000 flew overhead at 1,000 feet and dipped its wings in final salute.

President Kennedy's casket is off-loaded from Air Force One (Boeing VC-137C, SAM 26000) after his assassination in Dallas, Texas, in November 1963. Robert F. Kennedy and Jacqueline Kennedy are visible in the lift truck that has just lowered President Kennedy's casket to the waiting ambulance. A military honor guard salutes as the casket is moved toward the ambulance. *US Air Force photo*

Camelot to the Gipper

At the time of President Kennedy's death, jetliners were securing their hold on the airways of the nation as well as the world. Jets were replacing props in the air-refueling community, and the Boeing 707 was the best-known aircraft in the world. Perhaps the ultimate 707 during this time was SAM 26000, which served as Air Force One. This, of course, was the aircraft aboard which Lyndon B. Johnson was sworn into office as the thirty-sixth president of the United States.

Although SAM 26000 (designated VC-137C) carried special communications equipment, its interior furnishings were little different from those of the three VC-137A/B (Boeing 707-153) aircraft also assigned to Andrews Air Force Base. SAM 26000 was continuously backed up by the ubiquitous VC-137A/B known as *Queenie*, which had a crew of two pilots, a navigator, a flight engineer, and two communications operators, but usually carried only one communicator. SAM 26000 had the same number of flight crew but always carried two communicators, plus four security forces personnel and four stewards.

Johnson, who liked to call SAM 26000 "my own little plane," used a variety of aircraft to travel to his ranch near Austin, Texas. But whether it was a sleek Boeing 707 or a chugging, sputtering Convair, Johnson was a master at giving rides to those whose influence he needed. No other president was so adept at giving out the perk of a ride aboard Air Force One—or withholding it—but all presidents have used their magic carpet in the sky as a way of impressing and influencing other political leaders.

OPPOSITE: VC-137C (serial number 62-6000) sits on the ramp at Tempelhof Central Airport during President Reagan's visit to Berlin on June 12, 1987. A few days after this photo was taken, President Reagan gave his now-famous Brandenburg Gate speech with the Berlin Wall in the background, where he challenged the Soviet Union's Communist Party General Secretary Mikhail Gorbachev saying, "Mr. Gorbachev, tear down this wall!!" *Department of Defense, DF-ST-91-06394*

President Johnson made a state visit to Thailand on October 27 to 30, 1967. The president visited King Rama IX in Bangkok, with Air Force One operating from Don Muang Air Base, approximately 20 miles north of the capital. President Johnson later returned to Thailand on December 23, 1967, to visit troops at Khorat. *Pima Air and Space Museum*

John Trimble, a radio operator on Air Force One during LBJ's tenure, recalls how the president enjoyed the perks of his flying White House:

When talking on the radio, Johnson would lean back in his big chair, which had been specially installed and we had dubbed the throne, and talk with his subjects toiling far below. He could raise or lower the chair and desktop with the touch of a button. The [radio operator] was obliged to monitor these radio calls in order to take quick remedial action should it be necessary, as it frequently was. It was interesting listening, but I was always glad when the calls were over because I could catch up on some of my other duties. Johnson employed a mode of operation that he repeated over and over again. He would call a person to seek advice on a particular subject and infer that this person would be his only input prior to making an important decision. He would then talk with two or three different people on that subject using the same approach. I wonder if these "wise counselors" ever talked among themselves.

Johnson reconfigured the interior of SAM 26000. Additional seats were added, and the seats were reversed to face the rear of the aircraft—toward the president's compartment. Johnson liked to be able to keep an eye on his passengers, and the cherry wood partitions that separated the passengers from the stateroom were replaced with clear plastic dividers. President Johnson used SAM 26000 and other aircraft extensively on travels back and forth between Washington and his Texas ranch. He was also a world traveler and used the aircraft for his flights to Vietnam at the height of the war.

JOHNSON'S 118

For at least two months following the dark hours of the Kennedy assassination, Johnson refused to fly in Air Force One, for reasons unclear. At Andrews, rumors abounded, even the rumor that the president would begin flying via commercial airline.

But Johnson never stopped using the VC-140 JetStars and the VC-131H Samaritans that often took him to Texas. In fact, he used them so often that one of each was periodically stationed on a semipermanent basis at Bergstrom Air Force Base near Austin.

PRESIDENTIAL CONVAIRS: VC-131H SAMARITAN

Those who remember him say President Lyndon Johnson reveled in taking visitors to his Texas ranch aboard the Convair VC-131H Samaritan, a twin-engine, turboprop aircraft known in civilian jargon as a Convair 580. Looking down from his "king chair," a hydraulically operated seat that allowed Johnson—already a big man—to tower above his companions, the president regaled them with war stories and sometimes arm-twisted for a much-needed vote on Capitol Hill. A pilot from the era remembers, "Johnson was better than any other president at using a ride in an airplane as a perk for congressmen he wanted to impress. On those occasions when the VC-131H operated as Air Force One, LBJ invited senators and congressmen to travel with him and they became putty in his hands."

The 89th Aircraft Wing operated four VC-131Hs. The planes had originally been built as ordinary air force transports under the designation C-131D, powered by two 2,500-horsepower Pratt & Whitney R-2800-103W Double Wasp, twin-row, eighteen-cylinder reciprocating engines. The first C-131D (54-2806) made its final flight on July 28, 1954.

The US armed forces had used military versions of the Convair 380 airliner since the first MC-131A Samaritan medical evaluation aircraft (52-5781), first flown on March 5, 1954, and delivered on April 1, 1954. The first US Navy R4Y-1Z version (Buno. 140378) was delivered to Naval Air Station Anacostia, Washington, DC, in April 1955 for use by the assistant secretary of the navy. More than a dozen versions of the C-131 served in all branches of the US armed forces, which also operated T-29 (Convair 240) trainers and R4Y-2 (Convair 440) transports.

Allison T56-A-3 2,900-shp turboprop engines powered the sometimes-presidential VC-131Hs. On completing their 89th Airlift Wing duties, they were transferred to the District of Columbia Air National Guard. One of them (54-2816) suffered damage beyond repair when its landing gear accidentally retracted while on the ramp at Greenville, Texas, on May 12, 1977. The remaining three were transferred to a detachment of Naval Reserve squadron VR-52 in 1978. This unit was redesignated VR-48 in 1981. On November 15, 1985, three navy crewmembers lost their lives when one of the aircraft (54-2817) crashed on takeoff at Dothan, Alabama.

The operational career of the Convair C-131 ended on August 30, 1990, when VR-48 transferred the last aircraft (55-0299) to the Department of State. In its new civilian guise, the C-131H was assigned to Peru to support Peruvian National Police in drug interdiction activity.

Convair VC-131H (serial number 55-0299) is seen on April 24, 1982, years after it served the Johnson presidency. This aircraft was on standby at Bergstrom Air Force Base, the nearest large airport to the LBJ ranch outside of Austin, Texas. The aircraft was transferred to the US Navy, flew for the Department of State Air Wing, and was eventually sold surplus to a civilian operator. *J. G. Handelman*

OPPOSITE: President Lyndon B. Johnson traveled to Vietnam on board Air Force One to visit US troops on December 23, 1967. The president was returning to the United States after attending the funeral of Australian Prime Minister Harold Holt and meeting with other heads of state on December 21 and 22. En route home, Johnson stopped at Khorat, Thailand, and Cam Ranh Bay, Vietnam, where he was greeted by army Gen. William C. Westmoreland. Johnson's visit took place one month before the surprise Tet Offensive, which began on January 30, 1968. *National Museum of the US Air Force*

RIGHT: President Johnson (center) converses with Secretary of Defense Robert McNamara and Secretary of State Dean Rusk (right), as Bill Moyers (left) sleeps in a chair. The group is traveling on board Air Force One en route from Los Angeles to Washington, DC, on February 8, 1966. Johnson and his advisors were returning from meetings in Honolulu at Camp Smith, where they met with representatives of the Republic of Vietnam in attempts to further the peace process. *LBJ Library photo by Yoichi Okamoto*

Moreover, Johnson frequently asked Washington luminaries to join him as he continued to fly aboard the VC-118A, a military version of the Douglas DC-6 (53-3240), which his predecessor Kennedy had so thoroughly enjoyed. Johnson used the VC-118A on many local trips in the Northwest. A two-day October 1964 trip took the president to Teterboro, New Jersey, for a speech, then to Wilkes-Barre, Pennsylvania, then to New York City's LaGuardia Airport. The trip continued to Rochester, then Buffalo, then New York City, and finally back to Andrews. Johnson always made a point to be accompanied by as many people as possible, but on this trip, as on most, he was shadowed by his personal aide, Jack Valenti, a World War II B-25 Mitchell pilot later to be well

known as the chief lobbyist for the motion picture industry. Valenti told this author that Johnson, unlike Kennedy, had "no special attachment" to the VC-118A.

The VC-118A apparently ended its presidential duties in August 1967 when it was transferred to Europe. The aircraft was retired in June 1975 and is on display today at the Pima Air and Space Museum in Tucson, Arizona.

SAM 26000

As for the VC-137C, now better known to the public simply as Air Force One and to its users as SAM 26000, it was the ultimate version of the Boeing 707. It boasted a new, high-efficiency wing of 11 feet 7 inches greater span that earlier

LBJ'S FAVORITE: THE LOCKHEED C-140 JETSTAR

An aircraft displayed in front of the passenger terminal at Joint Base Andrews is a reminder of the era when the C-140 JetStar flew important missions for the air force. The JetStar was the first civilian business jet to be adopted for military use. The air force flew it from 1961 to 1989.

In the 1950s, the air force wanted jets to replace smaller prop-driven trainers and transports. One ideal candidate seemed to be the JetStar, designed by a Lockheed engineering team led by Kelly Johnson.

After looking at the JetStar on the drawing board, the air force selected it as a multiengine trainer and ordered ten planes, to be called T-40A. But in 1958, the service reversed this decision and ordered the more economical T-39A Sabreliner. The air force cancelled the T-40A contract, and none were ever built.

Soon afterward, the air force acquired five C-140A JetStars to be used to calibrate ground-based navigational aids in a mission known as "facilities checking." The first of these planes went to Scott Air Force Base, Illinois, in 1961. In later years, the planes operated at Clark Air Base, Philippines; Ramstein Air Base, Germany; and Richards-Gebaur Air Force Base, Missouri. The planes carried two pilots, a flight mechanic, and two enlisted technicians.

Subsequently, the service acquired five C-140B thirteen-seaters and six VC-140B JetStar eight-seat executive transports. The service quickly modified all eleven of these planes to the VC-140B standard. These were assigned to Andrews Air Force Base to provide transport to the nation's leaders. The planes carried two pilots and a steward. One VC-140B was used regularly in the 1960s to carry President Lyndon B. Johnson between Washington and his ranch in Texas. For a time, a VC-140B was stationed at Bergstrom Air Force Base near Austin, Texas, one of several aircraft (including a VC-131H Samaritan and Beech VC-6A) that became "Texans" during Johnson's tenure.

In later years, the facilities-checking C-140As were painted in camouflage and traveled as far as Vietnam to check navigational facilities. The facilities-checking mission remained in the air force until the 1990s, when it was taken over by the Federal Aviation Administration.

This aerial view of the LBJ Ranch shows a Lockheed JetStar on the ramp. Note the external power cart hooked up to the plane on the starboard side, near the nose. *US Air Force photo*

All air force JetStars were powered by four 3,000-pound-thrust Pratt & Whitney J60-P-5 turbojet engines, better known by their civilian term, JT12A-6. Typically, they had a cruising speed of 507 miles per hour.

Ironically, the airplane displayed at Andrews, although attired in military markings, isn't a military C-140 model at all. The display plane is a former civilian JetStar that never flew for the air force. A real VC-140B, one that was frequently used in the 1960s to transport Vice President Hubert Humphrey, is displayed at Wright-Patterson Air Force Base, Ohio. Another can be seen at the Pima Air and Space Museum in Tucson, Arizona.

LEFT: President and Mrs. Nixon board Air Force One for a trip to China on February 21, 1972. *Oliver F. Atkins/Official White House photo*

RIGHT: President and Mrs. Nixon share a moment together on Air Force One, en route to China. This photo shows the president's private quarters on board the VC-137C to advantage. *Oliver F. Atkins/Official White House photo*

models, giving the wing a new span of 145 feet 9 inches. The fuselage was 8 feet 5 inches longer than earlier 707s, for a total of 152 feet 11 inches. The new gross weight was 312,000 pounds. With its blue and white exterior and "United States of America" emblazoned on the fuselage, the aircraft seemed even more dazzling and glamorous than the 707s that were beginning to operate regularly on airline routes.

SAM 26000 set many point-to-point speed records flying from Andrews across the Atlantic. It was powered by the same 18,000-pound-thrust Pratt & Whitney

JT3D-3 turbofan engines that were belatedly retrofitted on its non-presidential cousins, the VC-137B models. The engines enabled Air Force One to reach a speed of 625 miles per hour. The four-member flight crew of pilot, copilot, navigator, and flight engineer sat forward of an eight-seat VIP compartment with galley and toilets. In the center fuselage was an airborne headquarters with conference tables, swivel chairs, a projection screen, and catnap bunks. The rear section had fourteen double reclining chairs, tables, galleys, and toilets.

As much as Johnson doted over SAM 26000, his successor seemed to take it for granted. When Richard Nixon became the twenty-seventh president on January 20, 1969, he began using the much-traveled Boeing 707 immediately. Nixon did not schmooze with the crew, however.

John Trimble, communications operator on the VC-137C, remembered, "LBJ was all over 'his' airplane. Nixon would give a wan little smile, greet the people within his immediate path, and disappear into his stateroom." Trimble recalls that presidential pilot Ralph Albertazzie said that during the five and a half years Albertazzie flew Nixon, the president never paid a visit to the flight deck.

WORLD TRAVELER

After the Johnson presidency, SAM 26000 continued its frequent travels. Nixon was in office less than a month when he made his first trip aboard SAM 26000, to Vietnam.

Shortly after Nixon's inauguration in January 1969, SAM 26000 went back to the Boeing factory for its first major overhaul. The aircraft was stripped to its metal shell from cockpit to tail. While engineers tested the aircraft's structure and systems, the interior layout was redesigned. The private quarters of the president were moved to the area forward of the wings, the most quiet and stable area of the aircraft. A staff compartment was built in the rear of SAM 26000.

Ironically, one feature of SAM 26000 that did not carry over into the Nixon administration was the taping system on board. By orders of the president, the system that recorded all incoming and outgoing calls on 26000 was removed.

In July 1969, President Nixon flew aboard Sam 26000 on a thirteen-day trip to six countries, culminating in a stop to meet the Apollo 11 crew in the middle of the Pacific Ocean.

In 1970, National Security Advisor Henry Kissinger used SAM 26000 to take him to the first of thirteen secret meetings with officials from North Vietnam. Flying these secret missions was a major undertaking. They were even kept secret from the secretary of defense, secretary of state, and director of the CIA. In 1971, Nixon gave 26000 an official name, *The Spirit of '76*, in honor of the coming bicentennial. A year later, the name would be transferred

ABOVE: This panoramic shot shows Air Force One's (serial number 62-6000) arrival in Peking, China, on February 22, 1972, with VC-137B (serial number 58-6970) in the background. President Nixon's visit was the first by a US president to communist China.
Official White House photo

OPPOSITE: President Nixon and White House Chief of Staff H. R. "Bob" Haldeman share a conversation in the office area outside the executive suite. Note the clock showing the time at three different locations, with Washington, DC, shown at far left.
Official White House photo

to the newly arrived second presidential aircraft, SAM 27000, but most continued to refer to it as Air Force One. In February 1972, SAM 26000 flew President Nixon on his historic visit to China, the first major step in normalizing relations with the world's most populous country.

ENTER SAM 27000

The arrival of SAM 27000 was a milestone. Beginning in 1972, it was no longer necessary for the VC-137B named *Queenie* to back up Nixon's flying White House. Now, the second presidential Boeing 707 came on the scene, wearing serial number 72-7000 and known for short as SAM 27000. Unlike

Queenie, the new VC-137C was meant from the start solely for presidential travel and served in that role from the beginning.

In fact, SAM 27000 replaced SAM 26000 as the primary aircraft for presidential travel, making it, in effect, the new Air Force One. Despite this change, the Nixon family preferred the interior layout of the older plane and used it whenever the family flew together.

SECURITY CONCERNS

The 1970s were a decade of hijackings and terror in the skies, and it was natural that experts would look for ways to protect Air Force One. Air force officials are loath to say much about it,

OPPOSITE: Both VC-137Bs (serial numbers 58-6971 and 58-6972) are shown on approach to land at Andrews Air Force Base. *Jim Hawkins collection*

RIGHT: VC-137C (serial number 62-7000) is shown shortly after its delivery to the air force. *Jim Hawkins collection*

BELOW: When President Kennedy was assassinated, Vice President Johnson was rushed to the nearest place the US Secret Service could ensure his safety—Air Force One. On September 5, 1975, Lynette "Squeaky" Fromme, a follower of convicted murderer Charles Manson and the Manson family, attempted to assassinate President Gerald R. Ford in Sacramento, California, with a Colt M1911 .45-caliber pistol. The gun had four rounds in the magazine, but the chamber was empty. Secret Service agents rushed the president back to Air Force One, which was parked at nearby McClellan Air Force Base. Defense Secretary Donald Rumsfeld can be seen on the airstairs while the president, aides, and Secret Service members wait to board Air Force One back to Washington, DC. *Photo by David Hume Kennerly; courtesy Gerald R. Ford Presidential Library*

but open sources disclose that both VC-137Cs were equipped with an infrared countermeasures self-defense system (ISDS) to protect the aircraft against heat-seeking missiles.

Developed from an earlier system called the AN/ALQ-144, ISDS was a fuselage- or pylon-mounted item designed to provide protection from antiaircraft missiles employing infrared homing devices. The system consisted of a multiband infrared countermeasures (IRCM) transmitter system, an electronic control unit, and an operator control unit. The system configuration varied according to the type of aircraft, but for most installations, it provided one transmitter per engine, with an associated electronic control unit. On the presidential VC-137Cs, the system was contained in rearward-facing fairings immediately above the rear of each of the planes' four engines.

Air force officers will not say how the defensive unit is manned or controlled, but according to literature published on the system, a single-operator control unit located in the cockpit controls one to four transmitters. The weight of the transmitter is 65 pounds, and the weight of the electronic control unit is 5 pounds. The operator control unit is a standard panel-mounted box measuring 146×125×57 millimeters and weighing about 2 pounds.

The system would be useful on any military aircraft, of course—the vision of terrorists or military irregulars setting up shop near an airport and popping off with shoulder-mounted, heat-seeking missiles is a traveler's nightmare—but it's apparently expensive and is employed in only about 150 turboprop and jet aircraft in production for US and Allied air forces.

ARMY HELICOPTER

While SAM 26000 received improvements for its role as Air Force One, the Marine Corps and army continued to share the mission of hauling the chief executive on short trips via helicopter. A marine officer told this author that Nixon "pretty much took the helicopters for granted" and showed no special interest in aviation or in the configuration of the aircraft. "But he was always amicable toward us, even when he appeared lost in thought on deep issues," the officer added.

By now, the army helicopters were part of an Executive Flight Detachment, commanded during Nixon's term by Lt. Col. Gene Boyer. He was also the pilot who flew Nixon from the White House to Andrews when the president resigned in August 1974. Copilot on that flight was Chief Warrant Officer 4 Carl Burhanan. Nixon made his last flight from the White House in Army One.

Although the crews and the cultures were separate, the Sikorsky VH-3A ("a blend of the civilian Sikorsky S-61 and the Navy SH-3A," according to a pilot) and Bell VH-1N Iroquois (Huey) helicopters operated by marines and army crews for the president were indistinguishable. All wore the same color scheme and the same words on the fuselage: "United States of America." Clint Downing, one of the army helicopter pilots,

told the author, "The only way you could distinguish which crew was flying was by their uniforms."

Downing adds, "During the time I was there, we [the army] shared the missions equally with the marines. We alternated trips, that is on one trip we would go ahead and preposition at the destination and travel with the president. The marines would pick him up on the lawn and take him to Andrews and then meet him upon his return. On the next trip this procedure would be reversed. The marines would travel ahead and we would pick him up from the lawn and meet him on the return.

"Army and marine VH-3s and the VH-1Ns were absolutely identical in every minute detail. Any changes to either or any aircraft was done by both army and marines in committee and had to approved by both. The only difference in the aircraft was the tail number [serial number].

"Another difference was that for backup support, i.e. maintenance, or extra passengers such as congressmen on trips, we used the army Boeing CH-47 Chinook and the marines used the Sikorsky CH-53."

ABOVE: An in-flight study of VC-137C (serial number 62-6000) during a flight to Andrews Air Force Base, Maryland. *US Air Force photo*

OPPOSITE: President Carter and his wife, Rosalynn, prepare to board Air Force One at Andrews Air Force Base on January 20, 1981, following the inauguration of President Ronald Reagan earlier in the day. Shortly before their departure, the fifty-two hostages held by the Iranians for 444 days were released. *Department of Defense photo by Staff Sgt. Prouty/DN-SC-83-02702*

"Those helicopters were navy aircraft and were provided in the navy budget," Downing remembers. "For a time, the military aide to the president was a navy admiral. The marines from HMX-1 [the presidential flying squadron, nicknamed the 'Nighthawks'] had a full-time liaison officer in the White House, and there was considerable pressure to give the entire mission to the marines. I have always believed that these factors were a big part of making the decision [in 1976] to close the Army Flight Detachment and turn the helicopter mission exclusively to the marines. Since the army has always been the major user of helicopters and has vastly more experience in rotary wing aircraft, I have always been of the opinion that this was a more appropriate job for the army. Apparently the Pentagon doesn't agree."

FORD AND CARTER

Gerald Ford became the nation's thirty-eighth president on August 9, 1974, when Nixon resigned over the Watergate scandal. After Army One flew him to Andrews on his last day in office, Nixon departed for California aboard SAM 27000. At the time of takeoff, the aircraft used the call sign Air Force One. The aircraft was approaching the Mississippi River when the hour of noon arrived and Gerald Ford's new job became official. (No president had resigned before, and none has since. The Constitution does not prescribe how it should occur, and the decision to make it happen at noon was Nixon's.) As the flight continued westward, the Federal Aviation Administration's Kansas City Air Control received the following radio transmission: "Kansas City this

Former presidents Ford, Nixon, and Carter are pictured on the steps of Boeing VC-137C (SAM 26000). Although three former presidents were on board, this flight was not designated Air Force One because the current chief executive was not on board. *US Air Force photo*

RIGHT: President Reagan joins the crew of Air Force One, sitting in the jump seat of the VC-137C. *US Air Force photo*

BELOW: A right front view of VC-137C (serial number 62-7000) parked on the tarmac at the Marco Polo Airport. The aircraft is standing by to fly President Ronald Reagan to Rome at the conclusion of the Venice Summit. *Defense Department photo by Gunnery Sgt. Hernandez/DM-ST-88-06835*

is former Air Force One, please change our call sign to SAM 27000." No longer president, Nixon continued his flight to California.

Ford's place in the history of presidential air travel consisted foremost of a bum rap: he sometimes misspoke and appeared awkward, so the press gave Ford a reputation for bumping his head on his helicopter doorway. In fact, it happened only once, and its significance was vastly overstated. In a famous utterance that has been quoted incorrectly by the press for decades, Lyndon Johnson had dismissed Ford as "a man who can't chew gum and fart at the same time," but though Ford was also photographed stumbling when boarding SAM 27000, he was, in fact, neither physically nor socially awkward. The reputation was unfair.

THE BEECH VC-6A SERVED THE PRESIDENT AND FIRST LADY

"Lady Bird's airplane" was the term used by President Johnson to identify the Beech VC-6A, military version of the King Air B90 with a special VIP interior (serial number 66-7943, constructor's number LJ-91). The air force acquired only one C-6, and it was never used for any purpose except to support the occupant of the White House.

Powered by two 550-horsepower Pratt & Whitney PT6A-20 turboprop engines and crewed by two pilots, the 6 to 8-passenger VC-6A had full pressurization for travel comfort at high altitude and all-weather navigation and deicing equipment.

The air force shelled out $436,000 to purchase the C-6 for the very specific purpose of meeting Johnson's needs. During the early part of its operational career, it was used to transport Johnson and members of his family between Bergstrom Air Force Base, Texas, near Austin, and the Johnson family ranch near Johnson City. During this time, the aircraft became informally known as the "Lady Bird Special." After leaving presidential service, the VC-6A continued in its special executive transport role with the 89th Airlift Wing until it was retired to the USAF Museum, to which it was flown on September 6, 1985.

The air force's sole Beech VC-6A (56-7943) was purchased expressly to meet Lyndon B. Johnson's needs and became "Lady Bird's airplane." It was one of several aircraft that spent considerable time at Bergstrom Air Force Base near Austin, Texas, servicing the Johnson ranch. Late in its career, the VC-6A doffed air force markings and national insignia and acquired the plain Jane color scheme shown here, making it difficult to distinguish from a civilian aircraft. *US Air Force*

While campaigning for the nation's highest office in 1976, Georgia governor Jimmy Carter traveled in a Boeing 727 dubbed *Peanut One*. After Ford's thirty-month tenure, Carter became the thirty-ninth president on January 20, 1977. Even before he took office, *Time* magazine wrote of his wish to make the presidency less regal: "He wants to minimize the use of Air Force One and to ride in an armored Ford LTD instead of the bigger and fancier Continental limousine most Presidents have used."

Although aircrews remembered Carter as gentlemanly and pleasant, he seemed in many ways an adversary to those who maintained and flew Air Force One. Several sources confirm that Carter exasperated radio operators by discussing classified material "in the open" on the airwaves—although this is very much a president's prerogative.

Carter ordered the *V* prefix on governmental aircraft designations removed, so SAM 27000, no longer a VC-137C, became merely a C-137C. The same change was made with other VIP aircraft at Andrews. Carter also ordered a toning-down of the markings on Air Force One and other VIP transports.

Carter was viewed as lively and exciting by crew members, especially those who had flown Nixon and Ford. On flights aboard SAM 27000, Carter sometimes danced in the aisles, sang favorite songs, or popped Bob Dylan into the music system.

Carter's paring-down of the visibility of Air Force One did not last long. Ronald Reagan undid all of it when he took office as the fortieth US president on January 20, 1981. Reagan also restored the *V* prefixes to all of the transports at Andrews.

President Reagan and his wife, Nancy, wave farewell as they prepare to depart for California aboard the C-137C Stratoliner following the couple's last day in the White House. *US Air Force photo*

OPPOSITE TOP: SAM 27000 is seen wearing "The Spirit of '76" titles on the nose of the aircraft and under the cockpit windows. *Jim Hawkins collection*

OPPOSITE BOTTOM: VC-137B (serial number 58-6970) taxies in after a flight at Andrews Air Force Base. *J. G. Handelman*

ABOVE: This left front view shows Air Force One arriving at Naval Air Station Bermuda with President George H. W. Bush aboard. He was attending a summit meeting with Prime Minister Margaret Thatcher of Great Britain on April 13, 1990. *US Navy photo by Photographer's Mate Second Class Doug Andrea*

LEFT: President George H. W. Bush disembarks from Air Force One at Naval Air Station Bermuda. The president and British Prime Minister Margaret Thatcher discussed making reductions to NATO's nuclear warhead stockpile as well as whether the Federal Republic of Germany should remain a full member of NATO. *US Navy Photo by Photographer's Mate Second Class Doug Andrea*

CHAPTER 5 FIVE

A New Presidential Aircraft

The president's plane was late. There was consternation in the nation's capital. "We're very concerned," said Mike Wallace, a civilian spokesman for the air force. The focus, fortunately, was not on any aircraft wending its way through the sky but rather on a long-anticipated "high-tech jet" (as the *Washington Times* called it) that was late showing up because it hadn't rolled out of the factory door quite yet.

The president's plane was still being built. It was June 1989, more than fifteen months after the "new, gee-whiz Air Force One" (the *Times*, again) was supposed to have been made available for Ronald Reagan. For five months there had been a new president, George H. W. Bush, who was sure to like the new plane. In fact, said one newspaper account, Bush, "a former naval pilot, will surely delight in its collection of frills and techno-toys." But not yet. There were going to be further delays, Wallace told the author of this volume. "It looks like we probably won't be able to deliver the new Air Force One until sometime in 1990," he said prophetically.

The secret behind the delay, or so the public was told, was a last-minute requirement that the electrical veins and arteries of the Boeing 747-200B, or VC-25A in air force jargon, be hardwired to resist electromagnetic pulse (EMP).

An invisible byproduct of a nuclear explosion, EMP can fry a computer chip, battery, or electrical wire not insulated against its onslaught. But insulation adds weight, and weight can slow down an aircraft or degrade its performance. Raise the weight of a new aircraft by too many pounds and it will not fly at all. In Wichita, Boeing's Carolyn Russell announced that the problem was "being addressed"—and it was. The finished VC-25A sacrificed weight elsewhere and came in at the original expected gross weight of 836,000 pounds.

Was the EMP issue a last-minute hang-up or merely the public excuse for a delay caused by many factors? A delay in meeting a military aircraft delivery schedule is hardly unusual, especially when the aircraft (in this case) has six million parts. The process of hardening some buildings, warships, and aircraft against EMP dated to the mid-1960s, and it strains credibility that EMP hardening was not part of the original requirement for the new Air Force One. More likely, the delay was simply a delay, a routine one interpreted with undue harshness by a press unversed in how planes are designed, developed, built, and fielded.

OPPOSITE: The aircraft that would be selected to replace the VC-137Cs as Air Force One was the Boeing 747-200, which in air force service was designated VC-25A. *Nicholas A. Veronico*

SELECTING THE NEW PLANE

The decision to pick a new plane for the president was difficult. Among those on Capitol Hill who keep their clutches on the purse strings, some feared that the public would see a new plane as wasteful and ostentatious. Others thought it would be a waste of money unless it was purchased in their district. But while the decision process was arduous, most in Washington took the choice of aircraft type for granted almost from the beginning.

It would be a 747 or nothing.

Among robust, wide-bodied jets with a "Made in USA" label, only the 747 was still in production at a cost and rate that made sense to a prospective buyer. The 747 was, after all, the standard against which every other large transport was measured, then and now.

RIGHT: Lockheed's L-1011 TriStar was presented as a possible Air Force One replacement, and the prototype, registered N1011, is seen here at after its first flight on November 16, 1970, at Palmdale, California. Lockheed built 250 TriStars between 1972 and 1984. One of the limiting factors of the L-1011 was its range (approximately 4,600 miles) in comparison to the 747-200 (approximately 6,075 miles). By the late 1980s when the next Air Force One was under consideration, the L-1011 was no longer being built, and the production line would have had to have been restarted for two aircraft. *Jon Proctor photo*

BELOW: Ronald Reagan served as the fortieth president of the United States. He was in office for two terms, from January 20, 1981, to January 20, 1989, and the selection of the VC-25A as the next aircraft to serve as Air Force One was made during his presidency. *White House photo*

When the 747 ushered in the wide-body era in the 1970s ("jumbo jets," the ads called them), three US companies were building large airliners. Boeing and Douglas were the best known. The third was Lockheed, maker of the L-1011 TriStar, an airliner with a good reputation for reliability that was less familiar to the public than the Boeing 747 or Douglas DC-10.

Lockheed had once been in the big league among airline manufacturers. The Constellation—a fetching four-engine, propeller-driven craft of the 1940s and 1950s—is regarded as one of the most graceful and majestic machines ever to take to the sky. But since production of the turboprop Electra ended in 1962, Lockheed had gone for years without building an airliner. Viewed in retrospect, the L-1011 TriStar seems to have been a kind of last gasp.

Developed in Palmdale, California, by a design team headed by William M. Hannan, the TriStar was the laggard among the trio of wide-body jets. The L-1011 made its first flight on November 17, 1970, a late debut compared with February 9, 1969, for the 747 and August 29, 1970, for the Douglas DC-10. Lockheed's wide-body airliner was launched just as the company faced crippling delays with its C-5 Galaxy transport for the air force. The TriStar itself was similarly delayed when its own engine configuration ran into technical problems. The L-1011 did not make its first revenue flight (wearing the blue and white colors of Eastern Air Lines) until April 26, 1972, a milestone that had been easily surpassed by the 747 on January 22, 1970, and the DC-10 on August 5, 1971.

TRISTAR POWER

The TriStar suffered teething troubles, principally with its 48,000-pound-thrust Rolls-Royce RB211-524 turbofan engines that had to be taken out of service for inspection at frequent intervals. But apart from these comparatively short-term problems, Lockheed's wide-body transport demonstrated efficient operation. It impressed operators, passengers, and the public with its reliability and with the low noise levels of the Rolls engines.

The L-1011 TriStar had a wingspan of 155 feet 4 inches, weighed 477,000 pounds when fully loaded, and offered a 19-foot 7-inch fuselage width, permitting an internal cabin width up to 18 feet 11 inches. There was nothing about the size and shape of this aircraft to make it anything other than a prime candidate for the president. In fact, one user in the Middle East had purchased an L-1011 TriStar and had spent more money plushing up the interior of the aircraft than the price tag of the plane itself. Moreover, there was a firm precedent for using the L-1011 as a military craft: Britain's Royal Air Force acquired nine of the planes for a dual mission as tankers and transports.

Lockheed eventually manufactured 250 TriStars. There was a short period when developmental problems with the DC-10 gave Lockheed an edge (and a slot in second place when competing with the 747), but it was a brief moment in the sun.

By the 1980s, when the Pentagon's requirement for a new presidential aircraft was taking serious shape, Lockheed was no longer building L-1011s. Foreshadowing a future in which competition would be even further reduced, the number of makers of jumbo jets had dropped to two.

PRESIDENTIAL CANDIDATE?

Lockheed's Jeff Rhodes remembers it a different way: "Did Lockheed want to propose an aircraft for Air Force One? No. Because the requirement called for a four-engine airplane. There was some talk at the time that McDonnell Douglas would submit a DC-10. But the specification that came out from the Air Force called for a four-engine aircraft, which kind of leveled the playing field. The only thing we could have offered at the time was a refurbished, redone L-1011 because they were no longer in production. It was supposed to be in time for Reagan to fly back to California, so that would have been [January 1989], but the L-1011 went out of production in 1982. So we wouldn't have had a new airplane."

Nevertheless, some in southern California—still in those days the center of the aviation industry, although its relentless decline had already begun—remember that Lockheed was serious about wanting to see Ronald Reagan in an L-1011.

"By the time the Air Force requirement gelled, we were no longer building L-1011s," remembers Frederick Newman, who worked at the manufacturer's Burbank, California, headquarters. "But there were airframes around with relatively few flying hours and we felt we could fix up a couple of them to meet presidential standards. We were not going to concede the playing field to the Boeing 747 without some effort on our part."

Apparently, the Pentagon reshaped the playing field with its four-engine requirement, and Lockheed's role never progressed beyond the planning stage.

DOUGLAS ENTRY

The story may have been slightly different in Long Beach, California, where Douglas Aircraft was one of the nation's premier builders of transports. In the mid-1980s, when a new aircraft for the president was being pondered, Douglas had very much retained its identity as the manufacturer of the famous DC series of airliners, including the DC-3, one of the most famous flying machines of all, and the DC-8, one of the first jetliners. Douglas's entry into the wide-body era, the DC-10 was in many ways as well known and as much admired as the 747.

Douglas viewed the historic advent of the 747 with caution. The southern California manufacturer, which had been acquired by McDonnell in 1966, forged ahead with the DC-10 with both American Airlines and United Airlines as launch customers. When the first DC-10 revenue flight was made by American on August 5, 1971, Douglas had taken just three years and four months between the decision to begin production and entry into scheduled service. This compares with three years and six months for the 747 (which came first because work was started earlier) and three years and nine months for the L-1011 TriStar.

The typical DC-10 airliner was powered by three 52,500-pound-thrust General Electric CF6-50C1 or -50C2 turbofans. The whale-shaped DC-10 was 182 feet 1 inch long with a wingspan of 165 feet 4½ inches. Maximum takeoff weight was 572,000 pounds. The wide-body seating was greeted with enthusiasm by travelers and caused one in-flight magazine to run an article called "A Revolution in How We Carry People."

continued on page 92

McDonnell Douglas Aircraft Company of Long Beach, California, proposed an Air Force One based on its commercial DC-10 and the KC-10 military transport/aerial refueling platform. The Air Force One paint scheme was applied to the DC-10 as seen in this illustration of the aircraft flying over Southern California. *McDonnell Douglas Aircraft Co. via Brian Baum collection*

McDonnell Douglas built a highly detailed model showing the interior configuration of its DC-10 Air Force One concept. The model shows conference rooms, the executive suite, galleys, various work areas, and the cockpit in great detail, down to the suggested finishes for each area. *McDonnell Douglas Aircraft Co. via Brian Baum collection*

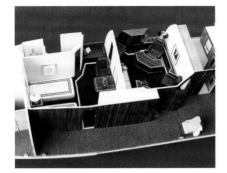
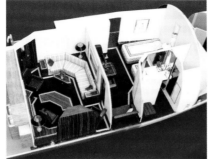

continued from page 89

Military DC-10

Any prospect that the DC-10 might be a candidate for presidential travel was enhanced when the air force ordered sixty military versions for a dual role as tanker/transports.

The Douglas KC-10A Extender tanker/transport, an off-the-shelf version of the DC-10-30CF (convertible freighter) airliner, was acquired to fill the air force's need—identified during Operation Nickel Grass, the US airlift to Israel during the October 1973 war—for a dual-role advanced tanker/cargo aircraft (ATCA). The United States' emphasis on rapid deployment forces in the late 1970s coincided with development of this long-range transport able to haul people and equipment while also functioning as a tanker. In the 1990s, the KC-10 fleet helped the USAF field an expeditionary force to deploy rapidly from US bases to Third World trouble spots.

The KC-10 is a genuine strategic asset with its capability to carry a full payload of 169,409 pounds over a range of 4,370 miles and its facility to be refueled in flight. The value of this tanker/transport is demonstrated when a fighter squadron deploys overseas, with the KC-10 carrying support equipment and personnel while also refueling the fighters en route.

For the tanker role, the lower section of the KC-10A fuselage is fitted with bladder fuel cells, increasing maximum usable fuel to 54,455 US gallons. A boom operator's station is located beneath the rear fuselage, where the operator sits in an aft-facing crew seat. A Douglas advanced aerial refueling boom (AARB) is located on the centerline further aft of the fuselage, with a refueling hose reel unit installed adjacently. This arrangement permits single-point refueling of USAF aircraft (with the boom) or US Navy/Marine aircraft (with the hose).

The first flight of a KC-10A took place on July 10, 1980, and its first air refueling on October 30, 1980, with a C-5 as the receiver aircraft. It was the following year that the air force began to ponder its new presidential aircraft. Douglas, now closely linked to the air force because of the KC-10 order, wanted the job. In Long Beach, California, Douglas proposed a luxury version of its DC-10 airliner.

ADVANTAGE 747

To understand why Boeing had such an advantage with the 747, we need to look at how this great aircraft transformed airline travel. So we now drop back to the mid-1960s, when the first jumbo jet was still a concept in the minds of officials at the Seattle company. Not everyone at that time was sold on the idea. When it first became clear that Boeing was working on a giant wide-body, turbo-fan-powered long-range airliner, scaremongers warned that an aircraft accident might now involve up to five hundred lives, while air traffic controllers celebrated an increased carrying capacity that would surely reduce the number of flights and make their jobs easier. The controllers' celebration overlooked a plain fact that was not yet solidly in evidence: the 747 almost single-handedly opened up air travel to everyone. Far from reducing the number of flights (though it caused a sharp reduction in the number of cross-country bus trips), the 747 was the force that brought millions into the fuselage and increased flights dramatically.

The 747 resulted from negotiations between Boeing's William Allen and Pan American World Airways' Juan Trippe. Pleased with the established track record of the Boeing 747, Trippe wanted a bigger aircraft that would save 30 to 35 percent per seat mile. This would bring lower fares and cargo rates, giving Pan American World Airways an edge over the competition. It is possible that he did not fully grasp that he was launching a revolution. On April 13, 1966, Trippe signed a contract for $525 million for twenty-five aircraft able to carry 350 to 400 passengers. This record order was a shot in the arm for Boeing, which was abruptly delivered from financial worries while free to launch an epoch-making aircraft.

At Boeing's new factory adjacent to Paine Field in Everett, Washington, the manufacturer devoted 14,000 hours of wind-tunnel testing to a variety of models, even after metal was cut and the first plane was being built. The company devoted 10 million engineering employee hours on the project. Far from being born overnight, the 747 underwent four years of continuous testing in areas ranging from metal selection to systems operation.

On February 9, 1969, with pilot Jack Waddell in command, the first 747 lifted into the sky over Seattle. Copilot Brien Wygle

An Air Force E-4B National Airborne Operations Center aircraft sits at the international airport in Bogota, Colombia, on October 2, 2007, waiting for Secretary of Defense Robert M. Gates. The VC-25A and E-4 are the two 747 types currently in the US Air Force's inventory. *US Air Force photo/Technical Sgt. Jerry Morrison*

and flight engineer Jess Wallick were also along on the maiden flight. Months later, after delays due to engine problems were resolved, Pan American's Najeeb Halaby, who replaced Trippe, used his own test-pilot credentials to personally wring out the 747 and pronounced it the safest, most comfortable, and most magnificently made plane in history.

Even before the crucial milestone of its first revenue flight on January 21, 1970, number crunchers wrote of the 747 in superlatives. When a 747 is fully pressurized, nearly a ton of air is added to its weight. A 747 has 6 million moving parts. Its tail height of 63 feet 8 inches is equivalent to a six-story building. The wing area of today's 747-400 model is 5,600 square feet, an area large enough to hold forty-five medium-sized automobiles.

In 1965, only 7 percent of Americans had traveled aboard an aircraft. By 1982, when the Pentagon was looking for a new Air Force One, the figure was rising toward 40 percent. Today, it is well above 75 percent. The wide-body revolution introduced by the 747, accompanied by the DC-10 and L-1011, transformed our lives as did no other event in aviation history. It's small wonder that polls consistently made the 747 one of the "most recognized" aircraft in history.

MILITARY 747

Boeing's Everett factory became the 747 Division of Boeing, with John Steiner named vice president for production development. The focus was centered on the airlines, but from the start, Boeing officials also saw military applications for the aircraft.

In 1973, Boeing proposed a version of the 747-200F to the air force for consideration as an advanced tanker/cargo aircraft (ATCA). To sell the proposal, Boeing modified the prototype 747 to perform dry aerial refueling linkups with a non-functioning flying boom apparatus. Boeing missed a chance for a lucrative 747 military order, however, when the air force chose the KC-10 Extender, the tanker-transport version of the Douglas DC-10, instead of the 747.

Boeing had better luck when the air force went shopping for a "Doomsday Plane." In 1973, the air force selected the 747-200 as its E-4A Advanced Airborne National Command Post (AABNCP). Subsequently known as the National Emergency Airborne Command Post (NEACP, pronounced "Kneecap"), the E-4A's job was to provide an aerial command center for the US leadership in wartime. For years, it was kept at Andrews Air Force Base on constant alert to carry the president or others in the chain of leadership known as the National Command Authority (NCA) during the initial hours or days of a general conflict. In the event of an attack on US soil, some leaders would be taken to an underground command post in Virginia while others would go aboard the E-4B to direct American forces.

Three E-4s (air force serial numbers 73-1676/1677 and 74-0787) were delivered in July 1973, October 1973, and October 1984. After a brief period with interim powerplants, they were powered by four General Electric F103-GE-100 turbofan engines. A fourth aircraft (75-0125) was delivered in August 1975 as the E-4B model with significantly improved avionics and communications equipment. The three earlier aircraft were subsequently brought up to E-4B standard. During the years when Jimmy Carter was president (1977–1981), their mission was de-emphasized, and they were transferred to the 55th Wing at Offutt Air Force Base, Nebraska, although they still appear frequently at Andrews.

Doomsday Plane

Modified extensively in the years after it was introduced, the E-4B was meant to accommodate the president (in his role as commander-in-chief of US forces) and key members of his battle staff on its vast main deck, partitioned into five operating compartments. These are the flight crew section, the NCA area (roughly a flying equivalent of the White House Situation Room), a conference room, battle staff area, and C3I (command, control, communications, and intelligence) area. A second deck provides a rest area for mission protocol.

This "war readiness aircraft" is equipped with nuclear thermal shielding, low-frequency/very-low-frequency (LF/VLF) radios, and extensive satellite communications equipment. Included is equipment to tie into commercial telephone and radio networks to broadcast emergency messages to the general population. The E-4B is distinguished from the original model by its super-high-frequency (SHF) system with antennas housed in a distinctive blister above and behind the flight deck. Every component of the aircraft, including engines, avionics, and wiring, was optimized for maximum flight duration. The E-4B's sustainability aloft is limited only by the oil lubricant of its engines.

The E-4B is now identified by the term National Airborne Operations Center (NAOC) and has been given new duties, including support in federal emergencies such as floods and forest fires. Some sources say that it is no longer needed in the doomsday role because of the advanced communications suite aboard the VC-25A (the current Air Force One), but at least one E-4B remains on some degree of alert to become the commander-in-chief's battle headquarters if necessary.

PRESIDENTIAL 747

In July 1986, the air force finally placed an order with Boeing for two presidential 747s. The White House had wanted the 747s on duty in time to carry the fortieth president, Ronald Reagan, home to California when Reagan left office in January 1989. Delivery of the airplanes was delayed, however, and the 747s did not carry a president until September 6, 1990, when George H. W. Bush was in office.

When searching for an aircraft to replace the VC-137Cs that would serve as Air Force One, Boeing suggested the 747SP (SP for Special Performance). The SP was built to fly long-haul routes up to 6,900 miles, typically from New York to the Middle East or South Africa, or San Francisco to Tokyo or Hong Kong, depending upon the customer. The 747SP was 47 feet shorter than a 747-100, -200, or -300, yet it had the same fuel capacity as the larger aircraft, giving it a range of 6,900 miles. By 1989, when the 747SP was offered as a potential Air Force One, engine technology was changing, and the aircraft was made obsolete by the forthcoming 747-400, which had more powerful and fuel-efficient engines.
Boeing via Brian Baum collection

OPPOSITE: The first of two presidential 747 (VC-25A) transports, in green configuration, is brought into the hangar at Wichita, Kansas, for outfitting to bring the aircraft up to standard for White House use. Note that the engine intakes are covered; the plane is being towed by a tractor. Both aircraft were manufactured in Everett, Washington, brought to Wichita for interior work, and returned to Everett for painting. *Boeing via Tom Kaminski*

LEFT: Back at the factory in Everett, home of all 747 models, a presidential 747 (VC-25A) is being transformed into Air Force One. A Boeing employee completes the finishing touches on decal letters, spelling *United States of America* on the side of the fuselage. Painting the aircraft in air force markings was the final step before delivery of the first plane in August 1990. *Boeing via Tom Kaminski*

RIGHT: This makes it official. Wearing safety harnesses as they work high up on the side of the airplane, Boeing employees apply the presidential seal decal to the new Air Force One. *Boeing via Tom Kaminski*

As mentioned previously, the delays in completing the two 747s received considerable attention in the late 1980s. The aircraft were built at the Boeing plant in Everett, Washington, and were flown in unfinished (green) configuration to the maker's facility in Wichita, Kansas. Here, the problems arose. The Associated Press attributed the problem to "doomsday technology"—-the hardening of internal wiring and systems against EMP—but the need to harden the aircraft against EMP had been seen from the beginning. The real reason for the delay was a more general problem integrating systems. By one account, this added $400 million to the original $261 million price for the two 747s, with the manufacturer picking up the tab.

Meanwhile, the press carped. The *Washington Times'* Frank Murray complained beneath the headline "Bush's high-jet jet delayed once again." The first of the two 747s was "already 15 months late," Murray grumbled in June 1989. The reporter wondered whether Bush would "ever get to use the new 747." Eventually, however, modifications to the first aircraft were completed, and it retraced the route from Wichita to Everett to be painted. The 89th Airlift Wing took delivery of the first craft on August 23, 1990. The second ship followed it on December 20, 1990.

The two aircraft (constructor's numbers 23824 and 23825) originally received air force serial numbers 86-8800 and 86-8900, reflecting their purchase in fiscal year 1986, but soon after delivery, they were renumbered 82-8000 and 92-8000 to be more in keeping with the numbers used on the VC-137s (62-6000 and 72-7000). This made it possible to give four successive presidential aircraft numbers that could be arranged in sequence: SAM 26000, 27000, 28000, and 29000. The aircraft also temporarily wore Federal Aviation Administration civil registry numbers while being flown by Boeing crews before being turned over to the Air Force.

As part of its preparation to receive the new (and delayed) flying White House, on January 20, 1989, the 89th Airlift Wing named Col. Robert C. "Danny" Barr presidential pilot, replacing Col. Robert E. Ruddick. It was no coincidence

The new Air Force One rolls out of the paint hangar in July 1990, ready at last to be turned over to the air force's 89th Airlift Wing. This is the first photo ever taken of the 747 as it appeared in presidential colors. *Boeing via Tom Kaminski*

SAM 28000 comes over the numbers and is about to touch down on Runway 30 at Long Beach, California, on March 18, 2009. Notice the large crowd of people in the background on hand to see the president's arrival. The VC-25A's paint scheme and highly polished undersurfaces can be seen in great detail from this view. Long Beach Airport is home to Douglas Aircraft/McDonnell Douglas Aircraft, which built the DC-8, DC-9, DC-10/KC-10, and MD-11 families of jetliners as well as the C-17 Globemaster III. Boeing acquired McDonnell Douglas in 1996. *Michael Carter/ Aero Pacific Images*

that this was inauguration day or that George H. W. Bush was taking office as the forty-first chief executive. While the president's chief pilot is chosen for experience, integrity, and ability, it is often left to the White House to choose the head of the presidential pilot's office. Barr and other pilots working with him became the first to receive training to fly the VC-25A, initially in a simulator at Everett. It is unclear, however, whether they were exposed to the actual aircraft while fitting work at Wichita was still taking place.

The Green Machine

At the end of March 1990 (five months before it was delivered to the air force), the first VC-25A, still green and sporting "N6005C" in a white rectangle on its fin, was conducting flight tests at Wichita. The aircraft was flying from the manufacturer's side of the airfield, which is shared with McConnell Air Force Base. No publicity attended these flight tests, and they received little notice in the press, although the occasional story could be found reporting that both planes were delayed.

The term "green" has come to refer to an airplane that has not yet been completed, but in this case, it had a literal meaning as well. In order to protect the polished silver aluminum covering the VC-25A, the airplane was covered with an alloy, a more pure aluminum that makes the airplane look green. The alloy covering essentially keeps the aluminum from oxidizing.

Among observers who witnessed the president's aircraft flying at Wichita were several dozen members of the F-4 Phantom Society, who were holding a reunion called a "Phancon" at McConnell. For the occasion, the air force side of the field was populated with F-4 Phantom fighters visiting from several units.

Jeff Rankin-Lowe, one of the participants, looked at the big green 747 with considerable interest. "It caught most of us by surprise," Lowe remembered a decade later, "as we were absorbed with a ramp full of 50 Phantoms and the metallic green color of the 747 was unexpected for many of us [who were] perhaps more used to zinc chromate primer. The speculation was that it was one of the 'Air Force One' aircraft, but no one knew for sure at the time." Another participant said he knew "exactly what it was" and believed civilian test pilots were doing the flying. "While we were there,"

the second participant said, "the aircraft remained on the [manufacturer's] ramp. It never once entered an Air Force part of the field except when necessary to land and take off."

The manufacturer's records list the first flight of the VC-25A as January 26, 1990, in Wichita, an apparent reference to a proving flight that took place nearly three years after the real first flight. The first VC-25A was formally accepted by the air force on August 23, 1990. It began flying President Bush almost immediately. The second aircraft was delivered on December 20, 1990.

QUEENIE TO PASTURE

In a ceremony at Andrews on June 14, 1993, the air force finally retired *Queenie*—the most presidential of almost-presidential aircraft—after nearly a quarter century of service. *Queenie* had joined the air force on June 12, 1959, as a VC-137A. The air force never earmarked *Queenie* for presidential duty, but she served as a backup to Air Force One (alias SAM 26000) for a decade. *Queenie* also carried Henry Kissinger to China to begin President Richard Nixon's dramatic rapprochement with that country in 1972. Moreover, *Queenie* was the first jet-propelled aircraft in the air force inventory specifically intended for the transport of personnel. Every previous jet had been designed to carry guns, bombs, or fuel.

Queenie carried US chief executives on numerous travels beginning on August 24, 1959, when on a European trip, Dwight D. Eisenhower became the first president to travel by jet. Presidents Kennedy, Johnson, Nixon, and Bush traveled on the aircraft. *Queenie* was retired to the Museum of Flight in Seattle, Washington.

SAM 26000 RETIRES TOO

While the new 747s were entering service, the air force continued to operate its first presidential Boeing 707, otherwise known as a SAM 26000. When state-of-the-art communications systems for the new aircraft were being developed, they were first tested on SAM 26000. In 1981, SAM 26000 carried former presidents Nixon, Ford, and Carter to the funeral of Egyptian President Anwar Sadat. In 1983, it carried Queen Elizabeth II on a visit to the west coast of the United States.

VC-137B (serial number 58-6970) makes a low pass over Boeing Field, Seattle, Washington, prior to its retirement to the Museum of Flight in 1996. The aircraft is on loan from the National Museum of the US Air Force and is open for tours, giving guests a first-hand look at the inside of a former Air Force One. *Brian Baum photo*

On May 20, 1998, after making presidential flights for more than three decades, presidential aircraft SAM 26000 made its final approach into Wright-Patterson Air Force Base, Ohio, to arrive at a permanent resting place at the US Air Force museum. The VC-137A was greeted by a crowd well-wishers at a museum ceremony, including Maj. Gen. Charles D. Metcalf, director of the museum; Air Force Materiel Command Commander Gen. George T. Babbitt; and several former crew members.

The aircraft that had served presidents from John F. Kennedy onward was replaced as the primary presidential aircraft by its stablemate SAM 27000 in 1972 and by the twin Boeing 747s in 1990. SAM 26000 continued to serve as a backup presidential aircraft until the day of its final flight. Its companion aircraft, SAM 27000, remained in service three years longer.

Retired Col. James Swindal, who piloted SAM 26000 through the Kennedy and Johnson administrations, was a special guest at the museum ceremony. Swindal said the most memorable moment aboard the presidential aircraft was a sad one—the flight back to Washington from Dallas after President Kennedy was assassinated.

The former Air Force One will continue to tell an important part of the air force story at the museum, where it is seen by more than one million visitors each year.

Boeing VC-137C (SAM 26000) lands following its final flight on May 20, 1998, at the National Museum of the United States Air Force, in Dayton, Ohio. SAM 26000 is the centerpiece of the museum's Presidential Aircraft Gallery.
US Air Force photo

Technical Description

The forty-third US president, George W. Bush, inherited an aircraft well known to the public, especially to movie audiences. Air Force One is the term by which the public recognizes the 747 transport (both of them) operated solely for presidential travel by the US Air Force's 89th Airlift Wing at Joint Base Andrews, Maryland. In actuality, Air Force One is the radio call sign for any air force plane carrying the president. For example, the call sign was used for a VC-9C Skytrain II, the military version of the Douglas DC-9, used by President Bill Clinton on one of his final trips as president, to Nebraska, on December 8, 2000. During his second term in office, Clinton also made a brief flight to Bosnia aboard a C-17A Globemaster III, which, for that brief interlude, became Air Force One. Of course, no president flew aboard more aircraft types than Lyndon Johnson (VC-137B/C, VC-131H, VC-140, VC-6A), and when he was aboard, each of these used the call sign as well.

Today even insiders use the term Air Force One in everyday conversation to refer to the pair of presidential 747s at Andrews. Designated VC-25A by the military, they are Boeing 747-200Bs or, in more technical terms, Boeing 747-2G4Bs.

OPPOSITE: A US Air Force VC-25 from the 89th Airlift Wing performs a touch-and-go practice landing at Atlantic City International Airport, New Jersey. Atlantic City International Airport hosts a wide variety of aircraft and is the home of the New Jersey Air National Guard's 177th Fighter Wing (the Jersey Devils) and US Coast Guard Air Station Atlantic City. The 89th Airlift Wing is located at Joint Base Andrews, Maryland. *US Air National Guard photo by Tech. Sgt. Matt Hecht*

A BIG PLANE

At the core of the air force's decision to purchase the 747 was a belief that no longer holds much currency. "In the 1980s when we wanted a new presidential aircraft, nobody would hear of anything with less than four engines," says a Pentagon officer. The age of twin-engine, two-pilot airliners had already arrived, but in the Pentagon, the air staff insisted that the occupant of the White House needed a jet with four engines and a flight crew consisting of no fewer than four: two pilots, a flight engineer, and a navigator. It was as if no progress had been made in aviation for a decade and extended twin-engine operations (ETOPS), the loosening of standards for long-distance flights, had never happened.

In the late 1960s, when every transport aircraft had at least three flight crew members, the 747 forged a unique reputation as the wide-body jetliner that had opened up air travel to the masses. The 747 first flew on February 9, 1969, and made its first passenger revenue flight with Pan American World Airways on January 22, 1970. By the time the air force went shopping, Boeing was close to the first flight of its two-pilot 747-400 (on April 26, 1988), which has since become the most numerous version of the famous jet, and was readily producing plenty of 767 airliners (first flight

on September 26, 1981), which could span vast distances not only with two pilots but with two engines. The air force was hearing none of it.

Douglas, however, heard the air staff loud and clear. Between 1978 and 1982, Douglas's Long Beach, California, plant manufactured sixty KC-10 Extender tanker/transports. The company aggressively promoted its three-engine DC-10 as the new Air Force One. But air force leaders had a tin ear toward any aircraft with fewer than four engines, and the marketing effort quickly sagged.

Four General Electric F103-GE-180 turbofan engines, better known by their civilian name, CF6-80C2B1, power the presidential 747s. The engines provide 56,750 pounds of thrust, about the same power as two railroad locomotives operating at maximum throttle.

The VC-25A has the standard 747 twin-lobe fuselage and is equipped to receive in-flight refueling via a standard air force–style boom receptacle. This receptacle slightly changes the shape of the extreme nose ahead of the cockpit. To ensure the VC-25A is self-sufficient on the ground, it is equipped with a second Garrett AiResearch GTC331-200 auxiliary power unit in the lower lobe.

LEFT: Air Force One (SAM 29000) is about to touch down on San Francisco International Airport's Runway 28 Left. When the president visits the City by the Bay, Air Force One will park on the airport's airside, and the staff will operate from the adjacent Coast Guard Air Station San Francisco. *Roger Cain photo*

RIGHT: Air Force One departs from Runway 28 Left with the coastal hills that form the spine of the San Francisco Peninsula as a backdrop. *Roger Cain photo*

This unique photo depicts both VC-25A Air Force Ones, serial numbers 82-8000 (at the rear) and 82-9000 (in the foreground), displayed for the public at the Joint Base Andrews open house in 2005. Both aircraft taxied out and took off, much to the crowd's delight.
Ken Kula

ON THE FLIGHT DECK

The cockpit of the VC-25A provides ample room for two pilots, a navigator, and a flight engineer and is thus substantially roomier than the flight deck of a comparable airliner.

On today's VC-25A, those crewmembers belong to the Presidential Airlift Group, known prior to April 1, 2001, as the Presidential Pilot's Office (PPO), with the status of an air force group under the 89th Airlift Wing at Andrews. The presidential pilot heads the group and reports to the commander of the 89th Airlift Wing. The next rung up the ladder is higher headquarters at 18th Air Force, located at Scott Air Force Base, Illinois. Going still higher, the 18th Air Force is part of Air Mobility Command, also located at Scott Air Force Base.

Whoever the presidential pilot happens to be, he looks down at the world from a lofty perch. On the VC-25A, the pilots sit on a flight deck situated 29 feet above the ground, roughly 100 feet in front of the main landing gear and 12 feet in front of the nose gear. Being this high up and this far forward demands careful thinking and a great deal of training when Air Force One is being maneuvered on the ground. Yet in spite of the enormous size of the VC-25A,

the flight deck is essentially the same size as the one on the earlier VC-137C.

In addition to the four flight crewmembers on the front deck, the VC-25A has crew positions for three airborne communications systems operators (ASCOs), although even on routine missions, it carries an extra ASCO for a total of four. In the early days of military flying, a separate crew position was needed on large aircraft to operate the radios, and the first airborne radio operators were drawn from the Army's Signal Corps. Since World War II, radio operators have typically begun their training at Keesler Field, Mississippi, and have been responsible for the high-frequency, very high-frequency, and ultra-high-frequency radios found on most transports. A 1982 study for the Pentagon's air staff by Chief Master Sgt. Ken Witkin, a radio operator on then–Vice President George H. W. Bush's aircraft, Air Force Two, changed the name of the career field from "airborne radio operator" to ACSO. The ASCOs who serve aboard Air Force One are part of the Presidential Airlift Group and serve under a chief of communications. Under this non-commissioned officer (NCO), the second slot is for a standardization and evaluation

ASCO. One source says that ACSOs serving on Air Force One are usually senior NCOs serving their final assignments before retirement. Some end up working as civilians for the White House Communications Agency, part of the Defense Information Systems Agency, which provides communications for the president and other staff members.

REACHING THE WORLD

The VC-25A has a special communications suite served by its three radio operator positions. The mission communications system (MCS) provides for worldwide transmission and reception of both normal and secure communications. The MCS includes multifrequency radios and eighty-five telephones for air-to-ground, air-to-air, and satellite communications. The airmen working at the radio stations have a huge responsibility for strategic communications, but they also handle prosaic tasks such as showing television and film programming to the president and other dignitaries. Much of the design work on the communications suite was directed not by a corporate executive or a senior officer but by one of the actual operators, Chief Master Sgt. Jimmy Bull, who at one time held the top communicator's job.

For more than a decade, the air force would not even acknowledge the location of the communications facility on the aircraft, which it now says is located on the upper deck behind the flight crew. Fifty percent greater in size than the suite on the previous presidential aircraft, the MCS suite has far more than the standard communications gear found on other big aircraft. Details must be speculative, since officials will not explain how communications for the president are set up or what equipment is provided for a doomsday scenario in which the commander in chief would be aloft at the outbreak of a war. Official sources will say only that the 747 carries a full suite of communications equipment, much of it installed by the former E-Systems (acquired in 1995 by Raytheon and operating as Raytheon Intelligence and Information Systems), enabling the president to talk to just about anyone. The communications gear effectively renders obsolete the air force's other Boeing 747 model, the E-4B National Airborne Command Post, operated by the 55th Wing at Offutt Air Force Base, Nebraska. With the communications afforded by the V-25A, the president no longer needs a separate command post in wartime.

It should be noted that in the event of a ballistic missile attack on the United States, Pentagon planning includes sufficient air-refueling tankers to keep Air Force One "tanked" with JP-8 aviation fuel indefinitely. It is believed that extra work was done on the VC-25A's engines to increase the amount of oil available to lubricate them, since this factor—coupled with crew fatigue—would limit the duration the aircraft can stay aloft. Some observers believe the president could be kept airborne for five or six days in the VC-25A with minimal difficulty.

STUDYING THE INTERIOR

Since the design features of the 747 are well known, it is the interior of the presidential VC-25A that makes Air Force One different. But do not expect a cutaway drawing to emerge from the manufacturer's public affairs shop. The air force not only won't reveal details of the interior of Air Force One (apart

OPPOSITE: The farthest-forward interior compartment of Air Force One (SAM 28000), looking forward, has seats along the fuselage wall. In many respects, the layout is similar to the first-class cabin of Boeing 747 airliners today. Boeing manufactured the VC-25A in Renton, Washington, but did the interior work in Wichita, Kansas. This interior view was released in 1990 and includes ashtrays generously scattered about. Soon afterward, the air force banned smoking on all of its aircraft, transforming the cigarettes that, until then, had been passed out to travelers on the president's plane into collectors' items. *Boeing via Tom Kaminski*

BELOW: The flying Oval Office—literally. As with every passenger seat on the aircraft, those shown here have first-class airline features, including seatbelts. But if the president chooses to stand while the aircraft is taxiing, taking off, or landing, no one is required to stop him. *Boeing via Tom Kaminski*

from a general description), but it also won't allow anyone to claim credit for any system or component of the aircraft. In late 2001, the company that made the MC-1 emergency oxygen mask, used by the VC-25A's flight crew, and by no other air force, ran an advertisement in *Armed Forces Journal* with a slick painting of the blue and white 747, proclaiming the firm's pride in being associated with the White House's executive jet. The company was requested—by whom is not clear—not to publish the ad again. Everyone associated with Air Force One takes great pride in the aircraft and its systems, but everyone does so in total silence.

Why is it so hard to learn about the interior of Air Force One? As one air force officer put it, "They now make a .50-caliber sniper rifle that is accurate to within a square inch at a distance of 3,000 yards. We don't want that sniper to know exactly where the president sits at his desk." On rare occasions when the press has been exposed to more than the rear passenger compartment, such as when *National Geographic*

magazine filmed a special on Air Force One, press activity had been closely watched. The air force public affairs establishment has tended to provide support (on rare occasion) to "puff pieces" about Air Force One, but not for serious analysis of the aircraft and its mission. The experience of Hollywood moviemakers provides a clue to how military bureaucracy and civilian media can clash. When Hollywood asked the air force for a guided tour of the inside of Air Force One, according to this same officer, "We said not only no, but hell no." However, months later, while the film was still in preparation, President Bill Clinton had dinner with Harrison Ford at the actor's ranch in Jackson Hole, Wyoming. In a conversational aside that sounded casual but wasn't, Ford asked for a tour of the VC-25A. Clinton instantly agreed to what the air force would not. The actor, film's director, and a handful of staff got their tour. No picture-taking was allowed, but as the officer remembers, "They had excellent memories. To a large extent, they accurately reproduced what they saw when they created the movie set."

LEFT: This area is known as the Annex and is situated close to the conference room. Seen here in executive configuration, it can be converted for use as an emergency medical facility.
Boeing via Tom Kaminski

RIGHT: This is another view of what was called the staff/secretarial area of the current Air Force One in the days before secretaries became associates and typewriters were replaced by laptops. The location of this area is on the right side of the VC-25A near the two-thirds point, just behind the conference room.
Boeing via Tom Kaminski

On the subject of movies, in *Independence Day*, Air Force One arrives at a base where little green men from another galaxy are preserved in glass bottles. And in *Escape from New York*, an especially unpleasant president (the late Donald Pleasance) cheats death by using an escape capsule in which the president can eject from the aircraft in an emergency. Air force officials will not confirm or deny the little green men, but they are adamant that the VC-25A does not have an escape capsule, nor are parachutes carried.

REAL-LIFE DRAMA

The real Air Force One has a copious interior, but it is still an aircraft with limited space, and the Hollywood version is bigger on the inside. The real aircraft includes a presidential suite comprising a conference/dining room, lounge/bedroom, and office space for senior members. A second conference room can be converted into a medical facility if required. The interior contains work and rest areas for a small presidential staff and a few media representatives as well as a rest area for the flight crew. In all, Air Force One provides seating for seventy passengers and twenty-three crewmembers.

The VC-25A is equipped with two complete galleys for food preparation, each capable of feeding fifty people. The aircraft has two self-contained airstairs, aimed at minimizing the need for ground support equipment and permitting the VC-25A to operate with minimal ground support facilities. The self-contained stair units provide entry and exit at a door located below and to the right of the main door, which is normally used for boarding using stairs provided at the base. A person boarding Air Force One via the airstairs would enter in the below-deck storage area and would climb an interior stairway to the main deck. The underfloor storage area is divided into general and special storage locations, the latter including sufficient food for two thousand meals. The lower deck also contains an automated, self-contained cargo loader and additional equipment.

Brig. Gen. James A. Hawkins, former commander of the 89th, remembers when he first saw a VC-25A during an earlier assignment in 1994: "I thought it was a remarkable aircraft. I compared it to the inside of a ship, a cruise ship, with the way it's proportioned, and with the mahogany in the conference room." Hawkins pointed out that while the VC-25A offers comfort suitable for the leader of the most powerful nation in the world, it is by no means opulent or ostentatious: "It is a practical aircraft."

DEFENDING THE VC-25A

The presidential 747 carries no armament. Secret Service agents escorting the president are armed, and their arsenal includes weapons rarely or never seen by the president, including sniper rifles and machine guns. Officials will not discuss the passive or active defensive systems aboard the aircraft, but the VC-25A does have infrared and radar warning devices, chaff, and flares. The pilots operate these from the flight deck.

Details on one such system have come to light. The VC-25A is equipped with the AN/ALQ-204 (MATADOR) infrared (IR) jammer, also known by the name Have Charcoal—the air

Technical Sgt. Albert Meriano III and Staff Sgt. Casey Watson, 1st Airlift Squadron flight attendants, prepare lunch at about 50,000 feet while in flight to Royal Air Force Mildenhall, England, on October 10, 2015. While a flight attendant's primary duties involve the safety of passengers, they are also well versed in the culinary arts and experts in customs regulations. *US Air Force photo by Senior Master Sgt. Kevin Wallace*

force's latest defensive weapon against heat-seeking missiles. (The word *Have* identifies the program as developed by Air Force Systems Command). The AN/ALQ-204 is part of a family of IR countermeasures systems designed for jet transport and providing coverage for the full 360-degree azimuth surrounding the aircraft. The system consists of multiple transmitters, a controller unit, and an operator's controller. The controller unit, which controls and monitors up to two transmitters, electronically synchronizes transmitters.

Each transmitter contains an IR source capability, which emits pulsed radiation to combat multi-IR-guided missiles. Preprogrammed multithreat jamming codes are provided, selectable on the operator's control unit, and all new codes can be entered as required to cope with new threats. (In addition to Air Force One, this defense against infrared missiles equips the Royal Air Force British Aerospace 146 aircraft of the Queen's flight and eighteen other head-of-state VIP aircrafts.)

Like many US military aircraft today, Air Force One has been fitted with a variety of infrared countermeasures to defeat heat-seeking missiles, including the ALQ-204, AAQ-24, and AAR-54 systems shown here. The largest potential threat to Air Force One is a lone attacker firing a shoulder-mounted, heat-seeking missile. *Ken Kula*

Four conformal antennas
portside mid-fuselage

Although hard to see, four conformal antennas have been added to the VC-25As on the port side at mid-fuselage above the words *of America*. These antennae enhance Air Force One's secure and non-secure communications capabilities. One antenna supports controller-pilot data link communications (CPDLC), which effectively replace high-frequency voice radio comms when the aircraft is over water. Air traffic control clearances can be sent directly to the aircraft, and the pilots can make requests back to controllers on the ground using this text-based system. The CPDLC system increases safety and makes communications more efficient and effective. *US Air Force photo by Senior Master Sgt. Kevin Wallace*

Supporting the AN/ALQ-204 is a Northrop-Grumman AN/AAQ-24 directional infrared countermeasures system that can detect multiple infrared guided missiles and jam them across a number of IR bands. Also fitted at the aircraft's tail cone is an AN/AAR-54 missile warning system, which detects ultraviolet emissions from an approaching missile's exhaust and can track multiple incoming sources. This information is then fed to a countermeasures system, such as a chaff/flare dispenser, to defeat a missile threat.

Maintenance on Air Force One is carried out by enlisted air force troops and not, as at many other air force bases, by air force or contract civilians. This is unusual, because the rest of

the 89th Airlift Wing now has shifted to contractor maintenance and uses civilians to keep its planes flying. Like Air Force One's cockpit crew and communicators, the maintenance people belong to the Presidential Airlift Group. The group also operates three C-20C Gulfstream III aircraft. About one thousand of the 89th Airlift Wing's six thousand members directly support VC-25A operations. The mission seems to be growing. During President Clinton's term (1993–2001), Air Force One touched down in forty-nine of the fifty US states and 112 countries. President Barack Obama flew 445 flights on Air Force One for a total of 2,799 hours and 6 minutes in the air, flying to all fifty US states and fifty-six countries.

CHAPTER 7 SEVEN

Additional Aircraft Serving the President

The United States has more than 15,000 airports across the nation, but only 5,054 are paved. The number of airports that can handle an aircraft the size of Air Force One is further reduced when considering the aircraft's weight, wingspan, and required turning radius. Figuring that a lightly loaded 747-200 (VC-25A) needs at least 7,000 feet of runway to operate safely, the number of Air Force One–capable airports is reduced to less than one thousand.

Having such a limited number of long runways in a country as vast as the United States where a president must travel to interact with his or her constituents requires alternate means of transportation. The Presidential Flight can operate a mini hub-and-spoke system, much like airlines do. Air Force One can be flown to a large airport, which acts as a hub, and the president can then be flown by helicopter to an off-airport site (the spoke). The alternative is to fly the president on one of the 89th Airlift Wing's VC-32s, which can operate from airfields shorter than 5,000 feet if configured for such a task.

The C-32 is a specially configured version of the Boeing 757-200 commercial extended-range jetliner. The C-32's primary customers are the vice president (using the call sign Air Force Two), the first lady, members of the Cabinet (such as the secretary of defense), and key congressional leaders. The C-32 replaced the C-137 and is flown by active-duty aircrews from the 89th Wing's 1st Airlift Squadron.

The C-32 body is identical to that of the Boeing 757-200 but has different interior furnishings and avionics. The passenger cabin is divided into four sections. The forward area has a communications center, galley, lavatory, and ten business-class seats. The second section is a fully enclosed stateroom for the use of the primary passenger. It includes a changing area, private lavatory, separate entertainment system, two first-class swivel seats, and a convertible divan seating three, which also folds out into a bed. The third section contains the conference and staff facility with eight business-class seats. The rear section of the cabin contains general seating with thirty-two business-class seats, a galley, two lavatories, and closets.

OPPOSITE: US President Donald Trump boards Air Force One prior to departure from Tri-State Airport in Huntington, West Virginia, August 3, 2017. *photo credit SAUL LOEB/AFP/Getty Images*

In a curious comment on the aircraft, an air force fact sheet states, "Because the C-32 is a high-standing aircraft, it is easier to see under and around it—an important security factor for protecting the plane and its passengers." The same release says that the C-32 "is more fuel efficient and has improved capabilities over its C-137 predecessor."

Well, almost. When the C-32 first reported on duty, the air force was embarrassed that it had shorter range than the ancient C-137. Range is an all-important consideration for people who travel frequently to Europe and do not want to stop on the way. The air force solved the problem by arranging to have Boeing Wichita install internal fuselage fuel tanks that increase range but reduce fuselage space. After the modifications, the C-32's 92,000-pound fuel capacity allows the aircraft to travel 5,500 nautical miles without refueling.

Compared to the C-137, the C-32 can operate on shorter runways down to 5,000 feet in length. The C-32 is equipped with a traffic collision avoidance system (TCAS) that gives advance warning of possible air crashes. Other items include

ABOVE: C-32 (serial number 99-0003) of the 89th Airlift Wing on the ramp at Andrews Air Force Base on October 9, 2004. The versatile C-32 carries the president, vice president, and members of the executive branch. *J. G. Handelman*

RIGHT: An airman from the Presidential Airlift Group prepares Air Force One for departure as a marine from Marine Corps Base Quantico stands at parade rest, awaiting departure protocols at Joint Base Andrews on July 15, 2015. President Barack Obama departed aboard an 89th Airlift Wing C-32 that day. *US Air Force photo by Senior Master Sgt. Kevin Wallace*

ABOVE: Members of Team Andrews greet President-elect Donald Trump as he arrives at Joint Base Andrews on January 20, 2017. Trump arrived in preparation for the 58th Presidential Inauguration that same day, when he took the oath of office. *US Air Force photo by Airman 1st Class Rustie Kramer*

TOP RIGHT: President-elect Donald Trump waves to a crowd with his wife, Melania Trump, as they deplane at Joint Base Andrews on January 19, 2017. *US Air Force photo by Airman 1st Class Gabrielle Spalding*

RIGHT: President-elect Trump salutes Col. Casey D. Eaton, 89th Airlift Wing commander, after deplaning from a C-32. Notice the presence of base security forces and US Secret Service members on the ramp and descending the airstairs, protecting the future chief executive and first lady. *US Air Force photo by Airman 1st Class Rustie Kramer*

a typical modern-day navigation system with GPS and a flight management/electronic flight instrument system. Inside the C-32, communications are paramount.

Boeing rolled out the first C-32 VIP transport in Seattle on January 30, 1998. The aircraft made its maiden flight from Renton Municipal Airport on February 11.

The USAF acquired four C-32 (Boeing 757-200) transports (98-0001/0004) plus two C-37As (Gulfstream V, 97-0400/0401) in the late 1990s. The mix of 757s and G-Vs was chosen after the air staff decided that it would be too ostentatious to purchase four Boeing 767s, the aircraft that was originally designated C-32.

The first of the four C-32s reached Andrews in 1998 following several delays that made the aircraft enter service about a year later than planned. The air force has a fifth 757-200 that is unmarked and used to transport special operations personnel. This aircraft is designated C-32B.

Beginning in 2017, President Donald Trump has made extensive use of the VC-32 as Air Force One, making numerous trips up and down the East Coast.

IN CASE OF EMERGENCY: THE C-20C

Inextricably linked to Air Force One and kept strictly out of the public's eye are the air force's three C-20C Gulfstream IVs, operated by the 89th Airlift Wing.

The C-20Cs, all of which serve at Andrews (air force serials 85-0049, 85-0050, and 86-0403), are emergency war-order aircraft designed to move high-ranking personnel quickly in the event of nuclear conflict and carries hardened, strategic communications equipment. The C-20C originally lacked the digital "glass" cockpit found on modern jets. Instead, the aircraft came equipped with analog instruments (round dials) that have changed little since the early days of aviation. The reason for the mechanical, rather than electronic, instruments is that the older gauges are invulnerable to

Aircraft mechanics and avionics technicians from the 89th Maintenance Group perform routine maintenance and inspections on a C-32A, C-37B, and C-37A from the 89th Airlift Wing at Joint Base Andrews. The 89th MXG maintains five aircraft types at the 89th, including C-32s, C-40Bs, C-37Bs, C-37As, and C-20Bs. A pair of C-20Bs and a C-32 are seen here on June 15, 2016. *US Air Force photo by Senior Master Sgt. Kevin Wallace*

LEFT: Secretary of Defense James Mattis (general, USMC, ret.) toured a number of high-technology companies in Silicon Valley in August 2017. His C-32 (serial number 99-0004), is seen on the ramp at Moffett Federal Airfield (formerly Moffett Naval Air Station) on August 11, 2017. *Nicholas A. Veronico*

BELOW: The C-32 that routinely carries the vice president is stocked with a variety of guest comfort items, including specially embossed playing cards. *Harry Lloyd collection*

EMP (electromagnetic pulse), the blast of energy that comes from a nuclear detonation and can fry anything that relies on electricity.

The air force has not disclosed details regarding the defensive system in the tail cone of the C-20C, which appears similar to systems on Air Force One. The system appears to be one known as IRCM self-defense system (ISDS), also known as AN/APQ-17 "Have Siren." This is designed to defend from attack from the rear hemisphere by IR guided missiles, primarily of the shoulder-fired type (which home in on engine exhausts). IR signals are emitted by the device, confusing and overloading the seeker head of an attacking missile, causing it to lose lock on the intended target. The C-20C has one such ISDS device in an aft fairing under the tailfin. The E-3 and Air Force One use ISDS, as do some fixed-wing turboprops and other jets.

The pilots and communications operators on the hush-hush C-20Cs are the same people who fly very prosaic and public Gulfstreams (C-20B and C-20H executive transports), but the C-20C version is a closely guarded secret. Although

public documents are available showing when they were built, fitted out, and delivered in the late 1980s, nothing has ever appeared in public about their purpose. When asked about the C-20C, officials said, "Our position is that we do not have any aircraft called a C-20C" (an Air Force Material Command program manager at Wright-Patterson Air Force Base, Ohio), or "No comment" (a communications operator at Andrews), or "You will have to ask somebody else" (a Gulfstream pilot). When the author requested information about the C-20C from Brig. Gen. Ron Rand, the air force's chief of public affairs, officials responded months later saying they would be unable to help. The following is informed speculation based on conversations with crews who discussed the hush-hush C-20C—but only a little.

Behind the curtain of secrecy is a program to ensure the survival of government leaders in the midst of a nuclear attack. The program is euphemistically referred to as the Senex program (derived from "senior executives"), and the C-20Cs are sometimes called Senex airplanes. Over the years, other terms applying to these aircraft have included the acronyms COOP (Continuity of [Government] Operations Program), COG (Continuity of Government), and PSSS (Presidential Successor Support System). These terms apparently apply not only to the mission of the C-20C but to war-readiness efforts in various locations around Washington, DC, including a now-defunct alternate underground facility for Congress in Greenbriar, West Virginia (abandoned after it was publicly revealed a few years ago), and an alternate National Military Command Center near Camp David, Maryland.

Apart from the presidential travel so familiar to the public, both Air Force One and the C-20C Gulfstream III have two additional functions. The first function is to assure the survival of the National Command Authority (NCA). Widely used incorrectly to refer to US military command

A US Air Force C-20B from the 89th Airlift Wing at Joint Base Andrews performs touch-and-go landings at Atlantic City International Airport, New Jersey, on April 16, 2013. Atlantic City IAP is the home of the 177th Fighter Wing of the New Jersey Air National Guard. The C-20B is used to transport members of the cabinet, congress, and defense department, as well as the first lady. *Air National Guard photo by Technical Sgt. Matt Hecht*

arrangements, NCA actually has a simpler meaning. It refers to the two officials authorized to release nuclear weapons—the president and the secretary of defense. The second function is to assure the survival of those in line for succession to the presidency, including the vice president, the speaker of the House of Representatives, and the president pro tempore of the Senate.

Whenever the president or secretary of defense travels, a C-20C Gulfstream IV always shadows their aircraft. If the president lands in Air Force One at, say, London's Heathrow Airport, a C-20C will land at nearby Royal Air Force Northolt

and remain on runway alert. Although it is clearly intended as backup transportation and as a source of communications support, an exact description of the C-20C's function remains elusive. When at home at Andrews, the C-20Cs have a special hangar near Air Force One's hangar and are rarely outside except when flying.

The C-20C is powered by two 11,400-pound Rolls-Royce F113-RR-100 (Spey Mk.511-8) turbofan engines. Because so little has been published about them, details on the three individual C-20Cs, beginning with the aircraft numbers assigned by their builder, include the following:

A 1st Airlift Squadron crew flies a mission to Europe on April 22, 2016. The 1st Airlift Squadron is part of the 89th Operations Group, which consists of two flying squadrons, the 1st and the 99th Airlift Squadrons, as well as the 89th Operations Support Squadron. The 89th Operations Group operates some of the most advanced commercial-based aircraft in the world, including the C-20B, C-37A, C-37B, C-32A, and C-40B (shown here). The 89th Airlift Wing maintains and operates Air Force One and fourteen other special air mission platforms. *US Air Force photo by Senior Master Sgt. Kevin Wallace*

All three of the C-20Cs (456, 458, 473) were delivered "green" from the factory to enable the installation of special mission equipment.

456 C-20C USAF 85-0050: Delivered March 15, 1985, to the 89th Airlift Wing at Andrews Air Force Base, Maryland. Fitted by E Systems. Now with 99th Airlift Squadron/89th Airlift Wing.

458 C-20C USAF 85-0049: Delivered March 28, 1985, to the 89th Airlift Wing at Andrews Air Force Base, Maryland. Fitted by E. Systems. Now with 99th Airlift Squadron/89th Airlift Wing.

473 C-20C USAF 86-0403: Delivered December 27, 1985, to the 89th Airlift Wing at Andrews Air Force Base, Maryland. Fitted at E Systems. Now with 99th Airlift Squadron/89th Airlift Wing. Observed at Shannon in overall gloss white with the same cheat line as SAM C-20, gold with blue outline, white color wraps under the nose and to the back of the aircraft, and under the engines. The underside is highly polished metal.

MARINE HELICOPTERS

On September 7, 1957, President Eisenhower was vacationing in Newport, Rhode Island, when his presence was required immediately at the White House. Typically, a return to Washington from Rhode Island called for an hour-long ferry ride across Narragansett Bay to the awaiting presidential airplane (then the *Columbine III*), followed by a 45-minute flight to Andrews and a 20-minute motorcade ride to the White House.

Ike directed his aide to find a way to get him to his airplane more quickly. The aide informed the president that a helicopter was on station in Rhode Island in case of emergency and could fly the president to the waiting plane. Eisenhower approved the idea and set a precedent with the seven-minute flight in a helicopter belonging to Marine Helicopter Squadron One (HMX-1) based at Marine Corps Air Base Quantico, Virginia, later to be nicknamed the "Nighthawks." The helicopter was a Sikorsky HUS-1 Seahorse.

Shortly thereafter, the president's naval aide asked HMX-1 to evaluate landing a helicopter on the South Lawn of the White House. The marines chose the Sikorsky HSS-1Z, which was virtually identical to the HUS-1 (both were redesignated VH-34D when the system for naming military aircraft was altered in 1962). Preliminary evaluation and test flights determined that there was ample room for safe landing and departure. Indeed, soon afterward, the air force performed similar landings with the Piasecki H-21B Workhorse and Bell H-13J.

By 1959, Eisenhower was routinely being transported between the White House and Andrews by marine and army helicopters. The air force continued to provide helicopters for possible emergency evacuation.

HMX-1 has continued the privilege of providing helicopter transport for the president to this day. The squadron is based at Marine Corps Air Base Quantico, Virginia, with a deployment of VIP helicopters stationed at Bolling Air Force Base (formerly Anacostia Naval Air Station) just down the Potomac River from the White House. For pilots flying with HMX-1, it is typically a four-year tour of duty flying Sikorsky VH-3D Sea Kings and VH-60Ns Seahawk helicopters. These specially marked executive transport helicopters are known as "white tops" for their distinctive white-over-green color scheme.

Marine helicopter pilots with 1,200 hours in the CH-53 and MH-53 are recruited from active duty squadrons, with twenty-five to thirty new pilots joining each year. Once on board HMX-1, pilots are given six to eight months of specialized training and often fly was copilot on Marine One missions. After two years with the squadron, pilots are selected as command pilots (flying the vice president, members of Congress, or members of the defense department/joint chiefs of staff) or Marine One pilots. HMX-1's commanding officer, executive officer, operations officer, and maintenance officer usually fly as pilots of Marine One, and their ranks are supplemented by other Marine One–qualified pilots who serve as copilot for the mission.

"During my time in HMX-1, I had about fifteen lifts with President Bill Clinton," said Frank "The Tank" Prokup (Lt. Col., USMC, ret.). "In the days after the September 11 attacks, I was a Vice Presidential Aircraft Commander and I flew Vice President Dick Cheney a lot. Most people don't realize this, but President

TOP: A photo from the roof of the White House shows VH-3D Marine One on the South Lawn. Note the circular discs under each of the helicopter's landing gears which are used to distribute the helicopter's weight across the soft lawn. *Courtesy Frank Prokup*

ABOVE: VH-3D Marine One departs the White House South Lawn carrying the president en route to Joint Base Andrews, where he will board Air Force One. *Courtesy Frank Prokup*

RIGHT: President George W. Bush returns to Washington, DC, late in the day on Tuesday, September 11, 2001. Marine One prepares to land on the South Lawn of the White House after a full day of travel for the president. Soon after touching down at the White House, the president addressed the nation reeling from the terrorist attacks that morning. *Photo by Paul Morse, courtesy of the George W. Bush Presidential Library*

March 17, 2014. This rarely seen view of the White House and the South Lawn was captured by White House photographer Pete Souza, who described the photo: "Returning from a visit to Walter Reed National Military Medical Center in Bethesda, Maryland, I asked the pilots of the Marine One helicopter if I could photograph from the cockpit as we approached the White House." The executive mansion is viewed under the pilot's arm as the VH-3D's throttles are manipulated on approach to landing. *Official White House photo by Pete Souza*

George W. Bush and Vice President Cheney were never in close proximity to each other in the weeks and months following the September 11 attack because of what was going on in the world. I would fly Vice President Cheney to Camp David, Maryland, on Sunday night or first thing Monday morning, and he would stay there through Friday, then we'd fly him home for the weekend. We did that week in, week out for all of September, October, and November 2001."

Flying the president or vice president sees the men and women of HMX-1 rotate on a two- or three-day tour depending on the needs of the executives. "All the duty pilots drive up to Bolling Air Force Base where the main briefings would be done, and if they had to reposition airplanes, they would be flown from Quantico up the Potomac River, past the Wilson bridge, land at Anacostia/Bolling, and you'd brief up there for the presidential mission," said Prokup. "Normally, the backup aircraft would be waiting at Anacostia, while the main helicopter, the one with the Marine One pilot and whoever the first officer (copilot) was going be that day, would fly over and land on the South Lawn of the White House a good thirty minutes before the pickup. Then two additional helicopters from Anacostia would be flying around

Washington, DC, while the president was picked up from the South Lawn of the White House. Once airborne from the White House lawn, two and sometimes three other identical helicopters will join in formation to create a shell game of sorts to make it harder to track from the ground which helicopter is actually carrying the president.

"It was done differently every time. It was probably an eight-minute flight from South Lawn over to Andrews Air Force Base where Air Force One is waiting, and at any given time, depending on the weather, it'd be about a forty- to fifty-minute flight up to Camp David."

As part of the aerial shell game, Marine One departs a location first and lands last at the next destination. That lets the president depart first, enabling the media to cover the chief executive's departure and the Secret Service to complete their tasks. These diverse groups then board white-top helicopters—or, depending upon departure point, the new V-22 Osprey tilt-rotors—that pass Marine One en route to the destination. The Secret Service deploys upon landing to assist those agents already on site, and the media gets into position to cover the president's arrival. "The president would depart first, then all of the press and Secret Service would get on the backup helicopters. We'd take off, chase 'em, pass 'em en route, land, they'd deplane, we'd shut down the helicopter, and they'd setup for the arrival of the president," Prokup said.

When returning to the White House, there are a number of approaches to the South Lawn, but none overfly the executive mansion. Coming from the area of the Capitol, Marine One flies down the Mall, then turns right prior to reaching the Washington Monument, overflies the Ellipse, and touches down on the South Lawn. Coming from the west, Marine One approaches the Mall over the Lincoln Memorial, flies east, then also turns to approach the South Lawn before reaching the Washington Monument.

"The doors on the VH-3 are on the port side, so you've got to bring the helicopter in with nose pointed at the White House. Pilots then slow to a hover twenty to thirty feet above the ground, do a right pedal turn (a pedal turn is where the main rotor stays at the same elevation but the fuselage rotates around

ABOVE: General Norman Schwarzkopf arrives at the US Naval Academy on board a white top HMX-1 VH-60N (Buno. 163263) on May 29, 2012. *J. G. Handelman*

OPPOSITE: Marine Helicopter Squadron One (HMX-1) departs Marine Corps Air Station Iwakuni, Japan, on May 27, 2016. President Obama visited MCAS Iwakuni and spoke with service members and their families after the Ise-Shima Summit. *US Marine Corps photo by Cpl. Nathan Wicks*

HMX-1 VH-3Ds, one of which is flying as Marine One, prepare to land at Joint Base Andrews as sunset approaches. For this trip, one of the C-32s will serve as Air Force One. *Courtesy Frank Prokup*

the axis of the rotor mast. This is done by changing the angle of the tail rotor blades, creating a yawing effect, allowing the fuselage to rotate in one direction. In this instance, the turn above the South Lawn is usually ninety degrees with the nose rotating to the right and the tail following around to the left), to position the port side door on the helicopter toward the residence. Once aligned, the copilot gives altitude cues to the pilot as the helicopter descends onto eight-foot diameter, one-quarter inch thick plywood circles anchored into the South Lawn," said Prokup. "These landings are very challenging. You're cueing the pilot when you're on approach, when he or she transitions into the hover, when the pedal turn begins, and during descent and touchdown. It was choreographed, and you probably flew at least a dozen times in practice with the

pilot you were going to fly with before you went and did it. This builds teamwork and trust among the flight crew."

The plywood circles distribute the helicopter's 20,000-pound weight across a greater surface area so that the landing gear does not sink into the soft lawn. This precise landing technique is practiced at Bolling Air Force Base.

When not flying government VIPs, HMX-1 pilots can also serve as White House Liaison Officers (WHLO, known as "Wheelos"), working directly with the White House staff and Secret Service on the road to support the president. WHLO go to all of the White House planning meetings and pre- and post-operation briefings and travel in advance of the president to coordinate helicopter travel when the chief executive is away from Washington, DC.

Marine One arrives at Joint Base Andrews, where the VC-25A Air Force One awaits the president of the United States. *Courtesy Frank Prokup*

The WHLO also coordinate the helicopter support for a detachment of armed marines who are usually out of sight but always near the president should additional security assets be needed. The presidential helicopter and those that support his movements require their own security forces for when the aircraft are on the ground, as well as maintainers to attend to the mechanical and avionics needs of Marine One and other white tops during deployments.

In addition to the squadron's VIP airlift duties, HMX-1 also supports the flying needs for the Officer Candidate School, known as OCS, at Marine Corps Base Quantico, Virginia, as well as operational test and evaluation of new aircraft and avionics for Marine Corps helicopters. HMX-1 recently evaluated the VH-71A version of the Eurocopter EH-101 three-engine, medium-lift helicopter as a replacement for the aging fleet of VH-3Ds.

DOUGLAS VC-9C SKYTRAIN

The Douglas Aircraft Company's VC-9C Skytrain, the military version of the commercial twin-engine DC-9-32 airliner, also served as Air Force One for three years, from 1992 to 1995. The aircraft's left cabin entry door carried the presidential seal when serving as Air Force One and had a slot for a placard for other high-ranking staff. First Lady Hillary Clinton used the VC-9C extensively during her tenure in the White House.

Retired air force pilot Ken Rice flew C-130 Hercules transports during the Vietnam War and was the pilot of the last US Air Force plane to receive battle damage during that conflict. Rice and his crew were participating in the evacuation of Saigon and were sitting on the ground when the final assault on the city took place. As they climbed out, their C-130 was hit by ground fire, and although the aircraft was damaged, they were able to complete the mission. After flying Hercs, Rice spent time as an aeronautical engineer working at the Air Force Weapons Laboratory at Kirtland Air Force Base, New Mexico, and returned to flying C-130s in the Pacific after a short stint with Braniff Airways before that company filed for bankruptcy. While flying C-130s from Japan and the Philippines, Rice applied for an interview with the 89th Airlift Wing. He was accepted and joined the squadron in 1985, flying C-140 JetStars and later the Gulfstream C-20s.

TOP: VH-3D (Buno. 159358) touches down on May 27, 2011, near the US Naval Academy at Annapolis, Maryland. Secretary of Defense Robert Gates was on hand to present the commencement address to the graduating class. *J. G. Handelman*

BOTTOM: VH-3D (Buno. 159354) passes above VH-3D (Buno. 159358) on May 26, 2017, as Vice President Mike Pence arrives to give the naval academy commencement address. Note that there are two VH-3Ds, one of which carries the vice president. The other serves as a backup as well as a decoy to confuse those on the ground as to where the vice president is located. *J. G. Handelman*

OPPOSITE: CH-46E (Buno. 153362) from HMX-1 departs the US Naval Academy on May 21, 2013. The CH-46s typically carries media covering the president and vice president. *J. G. Handelman*

TOP: HMX-1 MV-22B (Buno. 168332) supports a vice presidential flight on May 26, 2017. The MV-22 is replacing the CH-46 in the presidential airlift support role. *J. G. Handelman*

BOTTOM: An MV-22B Osprey from HMX-1 taxis on the runway at Williamsburg, Virginia, prior to takeoff. HMX-1 is responsible for transporting the president, vice president, cabinet members, and other dignitaries, as well as testing and evaluating new flight systems for Marine Corps helicopters. *US Navy photo by Mass Communication Specialist 1st Class Nathan Laird*

TOP: Queen Elizabeth II and Prince Philip stand with President and Mrs. Reagan during a ceremony to honor the queen's visit to the West Coast on February 23, 1983. There were rain showers up and down the California coast, and when the Queen arrived at the Santa Barbara Airport en route to the Reagan Ranch, the VC-9C was nosed into the hangar to enable welcoming ceremonies to proceed out of the weather. The president's limousine can be seen forward of the aircraft's nose. *Photo by Staff Sgt. Michael E. Lada*

BOTTOM: The executive interior of one of three VC-9Cs used by the 89th Airlift Wing to transport the president and other high-ranking government officials. *Courtesy Ken Rice*

Having flown the C-140, Rice was given the opportunity to select his next aircraft when the JetStars were retired. He chose the VC-9C. There were three of the type in the air force's inventory, only one of which, 682, was outfitted with a radio operator's station and secure communications capabilities. Rice flew all three VC-9Cs and accumulated more than 5,000 hours flying the type. He flew President George H. W. Bush as well as President Bill Clinton in the VC-9C Air Force Ones.

"Flying the president is exciting. The closest I could ever come to describing the mission was like flying into combat. The adrenaline starts flying as soon as the motorcade pulls up with all the police and the sirens, the flashing lights, and photographers and everything. It doesn't stop until the motorcade leaves when you get to where you're going," Rice said.

"It was more of a challenge for the pilots to keep the schedule right down to the second. We had a flight mechanic that sat in the seat between the pilots. On landing, he'd be

First Lady Hillary Clinton poses with Maj. Ken Rice (sixth from right) and crew. Photos with high-ranking guests were not routine, but Clinton gladly posed for this picture with Major Rice and crew after his last flight in command of the VC-9C just prior to his retirement. *Courtesy Ken Rice*

counting down to the second with the goal of setting the parking brake right on the hack of the very minute that we scheduled the flight. That was ideal. Normally, especially with President Clinton, you were never on time. He was always late. You were always rushing, trying to get back onto schedule.

"Our goal was to mark right to the second. The saying was, 'Fashionably late.' Which in our terms was one or two minutes, but never ever early. If you arrived early, there was going to be hell to pay. Mostly because of the people who were there to meet our passenger.

"I recall one time I took Vice President Dan Quayle into Chicago's Midway Airport. They have a busy airport so you really don't have a lot of control over your arrival. You're at the mercy of the controllers, and we had it all planned out to land on a specific runway and had estimated our taxi time as well as where we were going to turn off so we could pull into parking spot right on time. At the last minute, Chicago control switched us to different runway, and as soon as we pulled off the runway, we were right there at the parking spot, so we arrived almost eight minutes early. Chicago Mayor Richard Daley had not arrived to meet the vice president yet. After everybody left, Mayor Daley's staff was up on the airplane chewing my rear end for arriving early. So we tried never to do that."

Although the VC-9C's term as Air Force One was short, the aircraft performed admirably, and today all three of the type have been preserved in museums.

NAVY ONE

The US Navy was given the honor of flying President George W. Bush and having one of its aircraft serve as Navy One on May 1, 2003. President Bush became the first sitting chief executive to make an arrested landing on an aircraft carrier as well.

Trapping aboard USS *Abraham Lincoln* in Lockheed S-3B Viking of Sea Control Squadron Three Five (VS-35), President Bush flew out to meet the carrier as she returned from a ten-month deployment to the Arabian Gulf in support of Operation

Iraqi Freedom. The S-3B, Bureau of Aeronautics serial number 159387, was flown by Cmdr. John Lussier, commanding officer of the VS-35 Blue Wolves, with copilot Lt. Ryan Phillips giving up his seat for the commander in chief.

Once on board the *Abraham Lincoln*, President Bush addressed the nation, giving what became known as the "Mission Accomplished" speech—referring to a banner suspended from the island of the carrier.

Navy One was retired to the National Naval Aviation Museum in Pensacola, Florida, on July 17, 2003, having been flown to the museum by Capt. James P. Kelly, commander, Sea Control Wing Pacific. This presidential S-3B keeps company in the museum with a former HMX-1 VH-3A (Buno. 150613), which served presidents Nixon and Ford.

President George W. Bush successfully traps aboard the USS *Abraham Lincoln* (CVN-72) in an S-3B Viking (Buno. 159387) assigned to the Blue Wolves of Sea Control Squadron Three Five (VS-35) designated "NAVY 1." President Bush was the first sitting president to trap aboard an aircraft carrier at sea. The president met with the sailors as the *Lincoln* sailed home from a ten-month deployment to the Arabian Gulf in support of Operation Iraqi Freedom. Later in the day, he addressed the nation. *US Navy photos by Photographer's Mate Airman Gabriel Piper and Photographer's Mate 3rd Class Tyler J. Clements*

CHAPTER 8 EIGHT

Air Force One in the New Millennium

On one tragic day, the eyes of the world became focused on Air Force One. That day, Air Force One represented a safe haven for the United States' leader and a symbol of trust and confidence for people the world over that America's leadership was intact, there was continuation of the government, and that the president was safe.

That day was September 11, 2001.

The United States was under attack. Terrorists had hijacked four commercial jetliners that were scheduled to fly from the East Coast to destinations on the West Coast. The hijackings took place within the first hour of the flight while the aircraft were still full of fuel for the cross-country trip. The hijackers' first two targets were New York City's World Trade Center Towers in Lower Manhattan, which were followed later in the morning by an aircraft being flown into the side of the Pentagon in Arlington, Virginia, across the Potomac River from Washington, DC. The fourth and final hijacked airliner was intended to destroy the White House or Capitol building, but that attack was thwarted by passengers on board the jetliner.

The World Trade Center complex consisted of seven buildings, including the twin 110-story towers. The towers were the tallest buildings in the world when they officially opened on April 4, 1973. Known as the North Tower, One World Trade Center was 1,368 feet tall (417 meters) and located adjacent to Two World Trade Center, the South Tower, which was 1,362 feet tall (415.1 meters).

Less than a decade prior to the September 11 attacks, the North Tower had been a terrorist target on February 26, 1993, when a 1,336-pound truck bomb made from urea nitrate and hydrogen gas was detonated, killing six and injuring more than one thousand others. The bombers' intent was to topple the tower of One World Trade Center in hopes that it would collapse into Two World Trade Center, bringing both buildings crashing down. Although the truck bomb blew a 100-foot-diameter hole in One World Trade Center's underground parking area, the building remained stable. The two bombers and four accomplices were subsequently convicted in the attack.

OPPOSITE: On the morning of September 11, 2001, President George W. Bush was flown in Air Force One to secure locations in Louisiana and Nebraska as the attacks unfolded. The president's advisors and Secret Service detail did not want the president to return to the nation's capital until his security could be assured. *Robert Giroux/ Getty Images*

OPPOSITE: President George W. Bush participates in a reading demonstration the morning of September 11, 2001, at Emma E. Booker Elementary School in Sarasota, Florida. *Photo by Eric Draper, courtesy of the George W. Bush Presidential Library*

BELOW: President George W. Bush watches television coverage of the terrorist attacks on the World Trade Center on September 11, 2001, from his office aboard Air Force One. *Photo by Eric Draper, courtesy of the George W. Bush Presidential Library*

On September 10, 2001, President George W. Bush flew on board Air Force One to Jacksonville, Florida, where he kicked off his Putting Reading First initiative to provide $5 billion over five years to improve reading instruction in kindergarten through third grade. At the end of the business day, President Bush and his entourage departed Jacksonville and flew southeast across the state, landing at Sarasota-Bradenton International Airport. He overnighted at the Colony Beach and Tennis Resort on nearby Longboat Key.

On the morning of September 11, around 8 a.m., President Bush was given his daily situational briefing, and at that time no terrorist threats were reportedly communicated to the commander in chief. After the briefing, the president's motorcade departed for a prearranged visit to the Emma E. Booker Elementary School in Sarasota, where he would listen to, and read to, a group of students, followed by a photo opportunity and discussion of his education initiative.

Upon arriving at the school, President Bush exited the motorcade and entered the classroom, where he listened to six- and seven-year-old students reading from *The Pet Goat*. As the president entered the school, the first hijacked jetliner struck the North Tower at 8:46 a.m. President Bush was told that a small, twin-engine plane had struck the tower. Thinking it was a light plane accident, the president proceeded with the photo opportunity with the school children.

In reality, the aircraft that hit the North Tower was the hijacked 395,000-pound American Airlines Boeing 767-233ER, N334AA, that was intending to fly from Boston's Logan Airport to Los Angeles International Airport. Upon striking the tower, the aircraft's estimated 10,000 gallons of aviation fuel vaporized, killing its crew of eleven, eighty-one passengers, and many inside the tower.

At 9:03 a.m., United Airlines Flight 175, slated to also fly from Boston to Los Angeles with a crew of nine and fifty-six

continued on page 140

ABOVE: Having traveled from Sarasota, Florida, to Barksdale Air Force Base, Louisiana, on the morning of September 11, 2001, President Bush arrives midday at Offutt Air Force Base, Nebraska. Here he conferred with his top advisors and taped a message to the American people that was played while he flew on Air Force One back to Washington, DC, that afternoon. *Photo by Eric Draper, courtesy of the George W. Bush Presidential Library*

RIGHT: An F-16 escorts Air Force One late in the day of September 11, 2001, as the VC-25A flew from Offutt Air Force Base, Nebraska, to Andrews Air Force Base, Maryland. F-16s of the District of Columbia Air National Guard were supported by F-16s from the 119th Fighter Wing of the North Dakota Air National Guard as Air Force One neared the East Coast. *Photo by Eric Draper, courtesy of the George W. Bush Presidential Library*

continued from page 137

President Bush arrives at Andrews Air Force Base late in the day on September 11, 2001. *J. G. Handelman*

passengers, struck Two World Trade Center, the South Tower, between the seventy-eighth and eighty-fourth floors. Moments later, White House Chief of Staff Andrew Card walked up to the seated president and whispered into his ear, "A second plane hit the second tower. America is under attack."

Excusing himself from the gathered children a few minutes later, President Bush moved to an adjacent classroom where he was able to view TV coverage of the crashes and speak to Vice President Dick Cheney, National Security Advisor Condoleezza Rice, New York Governor George Pataki, and FBI Director Robert Mueller. Following the phone conversations, at 9:31 a.m., the president addressed the nation, saying, "Ladies and gentlemen, this is a difficult moment for America. . . . Today, we've had a national tragedy. Two airplanes have crashed into the World Trade Center in an apparent terrorist attack on our country."

Minutes later, at 9:37 a.m., American Airlines Flight 77 crashed into the Pentagon. En route from Washington-Dulles to Los Angeles, Boeing 757-233, N644AA, was carrying a crew of six and fifty-eight passengers, who all died instantly along with 125 Pentagon staffers.

By 9:43 a.m., President Bush had been driven through the streets of Sarasota and was back on board Air Force One. The VC-25A was off the ground eleven minutes later at 9:54 a.m.

At 9:59 a.m., the South Tower of the World Trade Center collapsed. It took nine seconds for the 110-story building to fall in on itself, taking another nine hundred lives inside the building and on the ground. Many who lost their lives were first responders attempting to climb more than seventy floors to help those injured in the catastrophe.

At 10:03 a.m., United Airlines Flight 93 dove into a field near Shanksville, Pennsylvania. Boeing 757-222, N591US, was traveling from Newark, New Jersey, to San Francisco, with seven crewmembers and thirty-seven passengers on board.

President Bush meets with guests aboard Air Force One en route to the funeral for Coretta Scott King in Georgia on February 7, 2006. Former President Bill Clinton, Sen. Hillary Rodham Clinton, Secretary of State Condoleezza Rice, Secretary of Housing and Urban Development Alphonso Jackson, and Jackson's wife, Marcia, talk with President Bush in the conference room on board the aircraft. *Photo by Eric Draper, courtesy of the George W. Bush Presidential Library*

These passengers were able to make calls from the aircraft to loved ones on the ground and found out about the hijacking conspiracy. With this information, the passengers bravely rushed the cockpit in the hopes of stopping their aircraft from being used as a missile to destroy another building, which was later determined to be either the White House or the Capitol in Washington, DC.

Less than thirty minutes later, at 10:28 a.m., the North Tower collapsed, killing an additional 1,600 people within the structure and on the ground below.

Colonel Mark Tillman, pilot and commander of Air Force One from 2001 to 2009, recalled the morning of September 11, 2001, in an October 1, 2014, Defense Media Activity video feature, saying, "I had to assume the worst, that the president would be under attack, so we got everything ready at the plane. There's numerous emergency action plans to take care of the president, so we had to go ahead and start putting those plans into effect to make sure everybody was ready to accept the president and get him where he needed to go.

"You had to accept that anything you were told was an actual threat against the president. Simple things like a man at the end of the runway with a video camera—you couldn't tell what he was or why he was there, but he came at the last second. We had to assume that was some type of plot against the president, so we countered it by taking off in the opposite

direction, climbing steep above the gentleman, and getting out of Sarasota rather quickly.

"We were not really sure what the plan was, so we headed out into the Gulf of Mexico to keep him safe. My thought process throughout the whole [event] was to get him to an air force base where there was high security; we could land and I would have all of the capabilities I ever needed."

At 11:45 am local time, Air Force One touched down at Barksdale Air Force Base, Louisiana, escorted by four F-16s from the 147th Fighter Wing of the Texas Air National Guard. After deplaning, President Bush made a number of phone calls to Vice President Dick Cheney, Secretary of Defense Donald Rumsfeld, and others. He then recorded a message to the American people that was be televised at 1:04 p.m. The speech lasted less than three minutes and was Bush's way of sending a message to the people of the world that its government was still in place and functioning.

Air Force One Flight Engineer Henry Frakes (Chief Master Sgt., ret.) recalled landing at Barksdale that fateful day: "Having started my air force career as a B-52 mechanic and spending five years crewing the E-4B National Airborne Operations Center version of the military 747 before coming to the Presidential Airlift Group, I knew the air force was well prepared to protect the president and by default, me, so I was more worried about my family back in Washington, DC., in case the attack escalated. When we dropped in on Barksdale and I watched a flight line full of B-52s taxi out as we landed, I truly thought things were escalating rapidly. What we didn't know at the time was that the Barksdale B-52 squadrons were in the middle of a pre-planned exercise."

Air Force One and its Texas Air National Guard F-16 escorts were back in the air by 1:37 p.m., this time headed for US Strategic Command (Stratcomm) headquarters at Offutt Air Force Base, Nebraska. The one-hour, thirteen-minute flight to Stratcomm headquarters was quick and uneventful. Here the president was taken to an underground command bunker, where he held a video conference with the National Security Council. At 4:33 p.m. local time, Air Force One was off the ground en route back to Andrews Air Force Base, Maryland. Nearing Maryland, Air Force One was joined in the sky by F-16s of the District of Columbia Air National Guard as well as F-16s from the 119th Fighter Wing of the North Dakota Air National Guard. After landing at Andrews, President Bush boarded Marine One for the short flight to the White House, touching down at 6:54 p.m. local time.

Having been safely shuttled across the midwest and eastern United States, a refreshed-looking President Bush addressed the nation at 8:30 p.m.: "I've directed the full resources of our intelligence and law enforcement communities to find those responsible and bring them to justice. We will make no distinction between the terrorists who committed these acts and those who harbor them." In spite of those reassuringly stern words, fear gripped the nation.

"On Air Force One, [the president] had all of his top [advisors] with him, and he had the ability to connect with any of the military leadership, which he did. Air Force One is a self-contained package, so he could make a lot of great decisions on board, and that's what he was doing . . . he was making great decisions for the country," Tillman said. "Our job on Air Force One was to make sure that the [command] on the ground was ready to [receive the president], keep him secure, hand him off to Marine One, and then keep our people and our aircraft secure as well.

"I'm getting a lot of credit for things that occurred on Air Force One, but it wasn't me. I was leading a group of people; there were three hundred folks working Air Force One, and more important, there were thousands of military that supported us, day in and day out, to [keep] the president safe. They are the ones that should be getting the big pat on the back because they did everything perfect to make it look seamless for all of us."

OPPOSITE: A US Navy SEAL team helps secure the airfield as Air Force One lands at Al Asad Airbase, Iraq, on September 3, 2007. President George W. Bush, Secretary of State Condoleezza Rice, Secretary of Defense Robert M. Gates, Chairman of the Joint Chiefs of Staff Gen. Peter Pace, US Central Command Commander Adm. William J. Fallon, Commander of Multinational Forces–Iraq Gen. David Petraeus, Commander of Multinational Corps–Iraq Lt. Gen. Ray Odierno, and others came to Al Asad to meet with Iraqi government leadership, sheiks from all over Anbar province, and US service members deployed to Iraq. *Defense Department photo by Cherie A. Thurlby*

TOP: Aboard Air Force One en route to Alabama, President Obama signs H.R.432, authorizing the Congressional Gold Medal to the "foot soldiers" who participated in Bloody Sunday, Turnaround Tuesday, or the final Selma to Montgomery Voting Rights March during March 1965. *Official White House photo by Pete Souza*

BOTTOM: Having arrived in Selma, Alabama, President Obama delivers remarks at the foot of the Edmund Pettus Bridge. *Official White House photo by Lawrence Jackson*

WHAT'S IT LIKE TO BE PART OF THE AIR FORCE ONE TEAM?

Henry Frakes served as one of the flight engineers on Air Force One from 2000 to 2010. He was there on September 11 and also made the secret trip with President Bush to visit US troops in Baghdad, Iraq, during Thanksgiving 2003. The retired chief master sergeant described his days with the Presidential Airlift Group:

"The standard avenue into the Presidential Airlift Group for flight crew used to be through one of the other squadrons at Andrews flying executive airlift with 757s, C-40s, Gulfstream IIIs/IVs/Vs/550s. Typically, a flight crewmember would spend three to five years flying those missions before they would be competitive for an interview with the Presidential Airlift Group. By that time, a prospect's reputation, how well he/she can handle the stress of the job, is well known. The mechanics of flying and maintaining a plane, prepping/serving meals, establishing communication channels, and building a flight plan can all be taught. Being able to function professionally under stress, readily owning up to and correcting your mistakes, and keeping your ego in check, are all valued traits that can't be taught and only come with maturity and experience.

"My hiring into the Presidential Pilots Office, later to be renamed Presidential Airlift Group, was not standard. I was the first flight crewmember to be hired from outside the 89th Airlift Wing at Andrews, thanks to Col. Mark Donnelly and Col. Mark Tillman. I already had five years' experience flying 747s out of Offutt with the National Airborne Operations Center program. Those 747s were significantly older than the two at Andrews, so I was well versed in handling real, in-flight malfunctions, not problems punched into a simulator. With four 747s at Offutt, I also spent a lot of time flight testing them after heavy maintenance, which is great experience for building systems knowledge and quality flight time.

"What is it like to fly the president? Initially, it's overwhelming. It took a year or two to get used to the fact that I was actually flying the president and his family around the world. I was hyper-conscious about every move I made and the risks inherent in some of my decisions. After a year or so, I gained more confidence and started to settle into a routine.

"Although it was an honor to fly three presidents, it was a privilege to have the opportunity to work with a group of aviators functioning at the top of their field. The Presidential Airlift Group's success was a team effort. Each crew position contributed to the success of the Air Force One operation.

"We typically carried a couple mechanics we called flying crew chiefs. There was nothing they couldn't fix and their skills weren't limited to airplanes. I've seen them fix everything from a de-icing truck to a forklift. They could be changing a starter control valve on one of the engines one minute then trying to get an ink stain out of a leather chair the next.

"Although our flight attendants' primary purpose was to keep our passengers safe, they were also dedicated to making sure the president, staff, and crew were well fed and comfortable. Unlike the other crew positions, the flight attendants were on their feet the entire flight. If they weren't prepping for a meal, they were cleaning up from one and catering to the staff's needs in between. The meals they prepared were all first class, made from scratch and were the product of a lot of planning, shopping, and cooking, often weeks in advance of a trip.

"Not all the airfields we landed at had the same level of security we're accustomed to in the United States. Our security team kept the aircraft safe from harm and were the gatekeepers of the manifest and anything carried onto the aircraft. While the rest of the crew might be enjoying a night off on the local economy, our security team diligently stood watch over the aircraft around the clock.

"The communication suite aboard Air Force One is what sets it apart from flying one of the airlines. Our communication team's forte was the ability to reach out to anyone, anywhere, at any time over secure or non-secure channels. They kept the president connected to the world in flight.

"Oftentimes, delays occurred that were beyond the crew's control. An event the president was attending might run long, a crisis might have emerged that needed to be addressed on the ground, or the weather may not be cooperating. Our navigators were masters at timing control, to the second. If we took off 30 minutes late, they would fine-tune our speed schedule, work out a shortcut, or find a pocket of favorable winds that would get the president back on schedule.

"The flight engineers were aircraft systems and performance experts and earned their keep managing the consequences of aircraft systems failures. When aircraft components failed, the engineers would determine what affect that failure would have on the flight and determine a way to safely keep the mission moving as scheduled until maintenance could be coordinated during a ground stop. They, and the crew chiefs, would put the aircraft to bed at night and wake it up in the morning. Oftentimes if ramp space was limited, they would taxi the jet to and from the arrival/departure location to streamline the preflight and post-flight process.

"Lastly, the pilots that flew the aircraft were exceptional. Like the other crew positions, most of the pilots were dual-qualified. In addition to being qualified in the 747, most maintained currency

in the 707, 737, 757, C-9, or Gulfstream. Besides being skilled pilots, what set these guys apart from most was their ability to manage risk under stress. Regardless of how well the crew is trained and the aircraft maintained, there was always something in the background working against the next on-time departure or arrival. Keeping the crew focused on recognizing and communicating those risks, then working as a team to mitigate them, was what they did best."

THE NEW 747-800 AIR FORCE ONE

In January 2009, the air force began looking to the future and initiated the process of replacing its VC-25As, at that time nearing the end of their second-third of the airframe's operational life. When constructed, the VC-25As were estimated to have a thirty-year life span, which was rapidly approaching. In theory, the VC-25As should complete their service to the president by 2020.

By March 2010, the air force had determined it would be cheaper to buy new aircraft rather than update its existing pair of VC-25As. The change in technology from the 1990s vintage VC-25As to a new aircraft based on Boeing's 747-800 would enable the air force to outfit the new Air Force One with the latest in communications equipment, defensive gear, and state-of-the-art, energy-efficient engines. The Under Secretary of Defense for Acquisition, Technology, and Logistics approved the Presidential Aircraft Recapitalization program's Material Development Decision in August 2012. In January 2015, the Presidential Aircraft Recapitalization program recommended Boeing 747-800s as the aircraft of choice to serve America's chief executive. At that point, the development of a new Air Force One, based on Boeing's ultimate version of the 747 jetliner, proceeded.

Two years later, everything was thrown into the air.

Prior to taking office, on December 6, 2017, President-elect Donald Trump tweeted, "Boeing is building a brand new

747 Air Force One for future presidents, but costs are out of control, more than $4 billion. Cancel order!"

The incoming president's social media message was a shot across the bow for the defense contracting industry, putting the group on notice to sharpen its pencils and tighten control over contract cost escalation.

Boeing responded later that afternoon with a statement, which read, "We are currently under contract for $170 million to help determine the capabilities of these complex military aircraft that serve the unique requirements of the President of the United States. We look forward to working with the US Air Force on subsequent phases of the program allowing us to deliver the best planes for the President at the best value for the American taxpayer." Nicely handled. The company's stock gained a little ground the next day, signaling that investors were confident that the planemaker could come to an agreement with the new administration.

Once the public back-and-forth was over, it was up to Boeing and the air force acquisitions team to deliver an Air Force One program at a greatly reduced cost. Following direction from the commander in chief, on August 4, 2017, less than seven months after Donald J. Trump took office, the secretary of the air force announced that Boeing had been awarded a single-source contract to replace the nearly thirty-year-old VC-25As that transport the president. The air force has negotiated a deal with Boeing for two 747-800s that had been built for the Russian airline Transaero.

The Russian carrier had ordered four 747-800s with General Electric Next Generation engines (GEnx-2B engines of 66,500 pounds thrust each) in November 2011, made the initial payments, and then entered bankruptcy in 2015. Two of the -800s built for Transaero were constructed as 747-85Ms (the -85M representing Dash 800 for the model, with 5M being the Boeing customer identification number). The two -85Ms are manufacturer's serial number (MSN) 42416 that was rolled out of Boeing's Everett, Washington, factory on May 12, 2015, and MSN 42417, which saw the sun for the first time in August 2015. Both were built and then sat in open storage at Everett while their fate was decided. With no customers on the immediate horizon, both -85Ms were flown to storage at Victorville, California, in February 2017.

In an air force news release, Maj. Gen. Duke Richardson, the Presidential Airlift Recapitalization program executive officer, said, "Purchasing these aircraft is a huge step toward replacing the aging VC-25As. This award keeps us on track to modify and test the aircraft to become presidential mission-ready by 2024."

Because the aircraft were bought using commercial contracts, the air force is not obligated to release the purchase price of the jetliners. The list price for a 747-800 is approximately $386 million, but it was disclosed that Transaero paid approximately $330 million per aircraft. It can be assumed that the air force acquired the jetliners at a deeply discounted price. Modification of the two 747-800s from airliners into the twin Air Force Ones is slated to begin in 2019, with the aircraft coming online in 2024.

The task of transporting the president by air has always been a military task and a military challenge. In today's world, it will continue to be so. The Boeing 747, alias VC-25A, and soon to be 747-800, known as Air Force One, is, in the end, not merely an ambassador but also a part of the nation's arsenal. In easy times and in difficult times, Air Force One serves America well.

Appendix I

PRIMARY AIR FORCE ONE AIRCRAFT (1943–PRESENT)

Aircraft	Serial/Registration	President(s) Served
Boeing 314 *Dixie Clipper*	NC18605	Roosevelt
Consolidated C-87A *Guess Where II*	41-24159	Roosevelt
Douglas VC-54C *Sacred Cow*	42-107451	Truman
Douglas VC-118 *Independence*	46-505	Truman
Lockheed VC-121A *Columbine II*	48-610	Eisenhower
Lockheed VC-121E *Columbine III*	53-7885	Eisenhower
Boeing VC-135	58-6971	Kennedy
Boeing VC-137C	62-6000	Reagan, Bush 41
Boeing VC-137C	62-7000	Kennedy, Johnson, Nixon, Ford, Carter, Reagan, Bush 41
Boeing VC-25A	82-8000	Bush 41, Clinton, Bush 43, Obama, Trump
Boeing VC-25A	82-9000	Bush 41, Clinton, Bush 43, Obama, Trump

After a short visit by President Obama to California's Silicon Valley, SAM 28000 lifts off from Moffett Federal Airfield's 9,197-foot long Runway 14L en route to the chief executive's next appointment in Los Angeles. *Nicholas A. Veronico photo*

Appendix II

AIR FORCE ONE: WHERE ARE THEY NOW?

Boeing 314A *Dixie Clipper* NC18605

Length: 106 feet

Wingspan: 152 feet

Height: 20 feet 4.25 inches

Powerplant: 4x 1,600-horsepower Wright Twin Cyclone radial engines

Maximum Speed: 210 mph

Range: 3,500 miles

Maximum Gross Weight: 84,000 pounds

Crew: 10

Passengers: 74

Disposition: *Dixie Clipper* was scrapped after World War II.

Ed Davies collection

Appendix II *continued*

Consolidated C-87A 41-24169 *Guess Where II*

Length: 66 feet 4 inches

Wingspan: 110 feet 0 inches

Height: 17 feet 11 inches

Powerplant: 4x 1,200-horsepower Pratt & Whitney R-1830-43 radial engines

Maximum Speed: 300 mph at 25,000 feet

Range: 3,300 miles at 188 mph at 10,000 feet

Maximum Gross Weight: 56,000 pounds

Crew: 4

Passengers: 25

Disposition: At the end of World War II, on October 30, 1945, *Guess Where II* was sent to the Reconstruction Finance Corporation's storage yard at Walnut Ridge, Arkansas. The transport was subsequently scrapped.

Douglas VC-54C 42-107451 *Sacred Cow*

Length: 93 feet 5 inches

Wingspan: 117 feet 6 inches

Height: 27 feet 7 inches

Powerplant: 4x 1,450-horsepower Pratt & Whitney R-2000 radial engines

Maximum Speed: 207 mph

Range: 4,200 miles

Maximum Gross Weight: 82,500 pounds

Crew: 7

Passengers: 15

Disposition: Displayed at the National Museum of the US Air Force, Dayton, Ohio

Photo courtesy National Museum of the US Air Force

Photo courtesy National Museum of the US Air Force

Photo by Mike Henniger

Photo courtesy National Museum of the US Air Force

Douglas VC-118 46-505 *Independence*

Length: 100 feet 7 inches
Wingspan: 117 feet 6 inches
Height: 28 feet 5 inches
Powerplant: 4x 2,100-horsepower Pratt & Whitney R-2800 radial engines
Maximum Speed: 360 mph
Range: 4,400 miles
Maximum Gross Weight: 107,000 pounds
Crew: 9
Passengers: 25
Disposition: Displayed at the National Museum of the US Air Force, Dayton, Ohio

Lockheed VC-121A 48-610 *Columbine II*

Length: 95 feet 2 inches
Wingspan: 123 feet 0 inches
Height: 22 feet 5 inches
Powerplant: 4x 2,200-horsepower Wright R-3350-75 radial engines
Maximum Speed: 330 mph
Range: 2,400 miles
Maximum Gross Weight: 107,000 pounds
Crew: 5
Passengers: 24
Disposition: Under restoration to fly by Karl D. Stoltzfus Sr., Dynamic Aviation, Bridgewater, Virginia

Lockheed VC-121E 53-7885 *Columbine III*

Length: 116 feet 2 inches
Wingspan: 126 feet 3 inches
Height: 24 feet 9 inches
Powerplant: 4x 3,400-horsepower Wright R-3350 radial engines
Maximum Speed: 330 mph
Range: 4,000 miles
Maximum Gross Weight: 133,000 pounds
Crew: 8
Passengers: 24
Disposition: National Museum of the US Air Force, Dayton, Ohio

Appendix II *continued*

Boeing VC-137B 58-6970, 58-6971, 58-6972

Length: 144 feet 6 inches

Wingspan: 130 feet 10 inches

Height: 41 feet 4 inches

Powerplant: 4x Pratt & Whitney JT3D-3B turbofan engines rated at 18,000 pounds thrust each

Maximum Speed: 530 mph

Range: 5,000 miles

Maximum Gross Weight: 258,000 pounds

Crew: 8-18

Passengers: 40

Disposition: 58-6970—Museum of Flight, Seattle, Washington; 58-6971—Pima Air and Space Museum, Tucson, Arizona; 58-6972—salvaged for parts at McConnell Air Force Base, Kansas, in 1998.

RIGHT: *Photo by Brian Baum;* BELOW: *Photo by Ron Strong*

Boeing VC-137C 62-6000 (26000) and 62-7000 (27000—*The Spirit of '76*)

Length: 152 feet 11 inches
Wingspan: 142 feet 5 inches
Height: 41 feet 4 inches
Powerplant: 4x Pratt & Whitney TF-33 (commercial designation: JT3D-3B) turbofan engines rated at 18,000 pounds thrust each
Maximum Speed: 600 mph
Range: 6,000 miles
Maximum Gross Weight: 316,000 pounds
Crew: 8
Passengers: 40
Disposition: 26000—National Museum of the US Air Force, Dayton, Ohio; 27000—Ronald Reagan Presidential Library and Museum, Simi Valley, California.

LEFT: *Photo courtesy National Museum of the US Air Force;* BELOW: *Photo by Nicholas A. Veronico*

TOP: VC-9C 73-1681 on display at the Castle Air Museum Atwater, California. *Photo by Ian Abbott*

BOTTOM: VC-9C 73-1683 on display at the Evergreen Air and Space Museum, McMinnville, Oregon. *Photo by Stewart Bailey*

OPPOSITE: VC-9C 73-1682 on approach to Dover Air Force Base, Delaware. *Photo Courtesy Roland Balik, 436th Airlift Wing/Public Affairs*

Douglas VC-9C 73-1681, 73-1682, 73-1683

Length: 119 feet 3 inches
Wingspan: 93 feet 5 inches
Height: 27 feet 6 inches
Powerplant: 2x Pratt & Whitney JT8D-9 turbofan engines rated at 14,500 pounds thrust each
Maximum Speed: 576 mph
Range: 2,900 miles
Maximum Gross Weight: 110,000 pounds
Crew: 13 (pilot, copilot, navigator, flight engineer, flight mechanic, four flight attendants, four security police, and depending upon the aircraft's configuration as radio operator for presidential communications)
Passengers: 42
Disposition: 73-1681—Castle Air Museum, Atwater, California; 73-1682—Air Mobility Command Museum, Dover, Delaware; 73-1683—Evergreen Air and Space Museum, McMinnville, Oregon.

Boeing VC-25A (747-200B), 82-8000 (28000), and 82-9000 (29000)

Length: 231 feet 10 inches
Wingspan: 195 feet 8 inches
Height: 63 feet 5 inches
Powerplant: 4x General Electric CF6-80C2B1 turbofan engines rated at 56,700 pounds each
Maximum Speed: Mach 0.84
Range: 6,735 miles
Maximum Gross Weight: 833,000 pounds
Crew: 26
Passengers: 102
Disposition: Both aircraft are currently in service.

Bibliography and Suggested Reading

Abbott, James A., and Elaine M. Rice. *Designing Camelot: The Kennedy White House Restoration.* New York: Van Nostrand Reinhold, 1998.

Boyer, Gene T., and Jackie Boor. *Inside the President's Helicopter: Reflections of a White House Senior Pilot.* Brule, Wisconsin: Cable Publishing. 2011.

Cross, James U., Denise Gamino, and Gary Rice. *Around the World with LBJ: My Wild Ride as Air Force One Pilot, White House Aide, and Personal Confidant.* Austin, Texas: University of Texas Press, 2008.

Graff, Garrett M. *Raven Rock: The Story of the U.S. Government's Secret Plan to Save Itself—While the Rest of Us Die.* New York: Simon and Schuster. 2017.

Haigh Sr., John L. *Air Force One: An Honor, Privilege, and Pleasure to Serve.* Tarentm, Pennsylvania: Word Association Publishers, 2013.

Hardesty, Von. *Air Force One: The Aircraft that Shaped the Modern Presidency.* Chanhassen, Minnesota: Northword Press, 2003.

Johnson, Lyndon Baines. *The Vantage Point: Perspectives of the Presidency, 1963–1969.* New York: Holt, Rinehart, & Winston, 1971.

L'Heureux, Ray, and Lee Kelley. *Inside Marine One: Four U.S. Presidents, One Proud Marine, and the World's Most Amazing Helicopter.* New York: St. Martin's Press, 2014.

TerHorst, Jerald, and Ralph Albertazzie. *The Flying White House: The Story of Air Force One.* New York: Coward, McCann & Geoghegan, 1979.

Veronico, Nicholas A., and Jim Dunn. *21st Century U.S. Airpower.* Osceola, Wisconsin: Zenith Press, 2004.

Walsh, Kenneth T. *Air Force One: A History of the Presidents and Their Planes.* New York: Hyperion, 2003.

Index

Index *continued*

Index *continued*